The Challenge of Climate Change

The Challenge of Climate Change: Which Way Now?

By

Daniel D. Perlmutter
University of Pennsylvania

and

Robert L. Rothstein
Colgate University

WILEY-BLACKWELL
A John Wiley & Sons, Ltd., Publication

Blackwell Publishing was acquired by John Wiley & Sons in February 2007. Blackwell's publishing program has been merged with Wiley's global Scientific, Technical and Medical business to form Wiley-Blackwell.

Registered office: John Wiley & Sons Ltd, The Atrium, Southern Gate, Chichester, West Sussex, PO19 8SQ, UK

Editorial offices: 9600 Garsington Road, Oxford, OX4 2DQ, UK
The Atrium, Southern Gate, Chichester, West Sussex, PO19 8SQ, UK
111 River Street, Hoboken, NJ 07030-5774, USA

For details of our global editorial offices, for customer services and for information about how to apply for permission to reuse the copyright material in this book please see our website at www.wiley.com/wiley-blackwell

Library of Congress Cataloging-in-Publication Data
Perlmutter, Daniel D.
The challenge of climate change : which way now? / by Daniel D. Perlmutter and Robert L. Rothstein.
p. cm.
Includes bibliographical references and index.
ISBN 978-0-470-65498-9 – ISBN 978-0-470-65497-2 1. Climatic changes. 2. Global warming. I. Rothstein, Robert L. II. Title.
QC903.P445 2010
363.738'74–dc22

2010022400

ISBN: 978-0-4706-5498-9 (hbk) 978-0-4706-5497-2 (pbk)

A catalogue record for this book is available from the British Library.

This book is published in the following electronic formats: eBook 9781444328530; Wiley Online Library 9781444328523

Set in 10.5 on 12.5 pt Times by Toppan Best-set Premedia Limited
Printed and Bound in Singapore by Markono Print Media Pte Ltd

1 2011

Daniel D. Perlmutter

To my wife, Felice, and to my grandchildren Seamus, Noa, and Lev, with hope for a better future for all grandchildren.

Robert L. Rothstein

To my Beloved wife Jane – an inspiration to all environmentalists!

Contents

Preface

There is a recurrent theme that appears in science fiction stories concerning how humans might respond to an invasion by aliens from some extraterrestrial source. The men and women in the tales are usually divided in their responses, siding as might be expected with their particular national or ethnic identities, their economic interests, or the degree of their xenophobic tendencies. This remains the case as long as the invaders are friendly or perhaps even indifferent to human aspirations; however, the scenario changes radically if it is found that the extraterrestrials are hostile to *Homo sapiens*.

It then transpires that all humanity unites in the face of danger from the outside. The serious threat to us all redefines what we understand to be the "other." Given this new attitude among earthlings, the stories usually have a happy ending. By cooperative resistance from humanity the enemy is defeated, or at the very least gives up its idea of conquest and leaves us in peace, planning for an optimistic future.

As a suitable allegory, this story might be useful in considering the threats to our planet that arise from global warming, as well as our responses. As the scientific evidence for the effects of greenhouse gas emissions to the atmosphere becomes increasing undeniable, it is clear that the damages from global warming will be widespread over the planet, rather than confined to a few locations at the poles. At this moment in our relatively short history of dealing with this "invasion," each nation and each economic interest group is looking for ways to come out ahead. The developing nations, especially China and India, do not want to accept any policy that would slow down their economic development. The economic interests that benefit in a big way from combustion of fossil fuels initially denied that there was any problem at all, arguing that the scientific evidence was not convincing; later, when total denial was no longer an acceptable posture, they began to invest some of their vast resources in alternative energy sources. The result is confusion in the public mind, uncertainty on the extent of the threat, and a lack of unity that might otherwise pressure governments to take stronger action. With rather few exceptions, voters are not prepared to sacrifice any significant part of their comforts to counter some as yet invisible hazard.

When will the story change? In our allegoric tale it was necessary to wait until the threat was clearly visible to all, when intergenerational survival seemed more important than short-term gain, and when it was universally recognized that the threat could only be handled by immediate world-wide cooperative action. It is our hope that by contributing to a process of wider consciousness raising, this book will contribute to making possible the same recognition with respect to global climate change.

But is another book on the problems of global warming really necessary? We are inundated virtually every day by a mountain of books, articles, and reports seeking to enlighten us about the dangers of global warming and what we – all of us, individuals, states, international institutions – need to do to avert the worst dangers and to adapt to what we cannot avert. Indeed many of these works are first-rate endeavors by eminent scientists and environmental activists. Is this book merely bringing more coals to Newcastle, or do we have something of value to contribute?

We do believe our cooperative enterprise has produced something of value, something that is different from the usual analyses or exhortations to the unconvinced. We started out with a common interest in environmental issues – how could one not be interested? – but with two completely different disciplinary backgrounds, one of us a Professor of Chemical Engineering, the other a Professor of International Relations. Initially each of us focused on what we knew best, then we sought convergence on a range of policy choices in different time periods. An easy fit emerged from the melding of our individual experiences, perspectives, and emphases.

The results are presented in the concluding chapters to provide a list of possible policies, choices based on characteristics that are scientifically, politically, and morally sustainable. More importantly, specific priorities are suggested for the short term, the medium period, and the long run. This is not intended to be an original scholarly contribution to either engineering technology or international relations, but rather a set of recommended approaches for non-expert readers that might help them find their own positions on some of the most vital issues of the day.

Finally, it should be noted that we write in the aftermath of the Copenhagen conference of December, 2009. Some have billed this conference in nearly apocalyptic terms, as mankind's last (or best) chance to keep global temperatures from rising to unsustainable levels. We have a different perspective. We believe global conferences are not the best setting to establish meaningful targets and timetables for rapid and resolute policy initiatives by the major contributors to global warming. But more to the point, what came out of Copenhagen – a weak compromise document lacking essential details on implementation – left unchanged the need for sustained and increasingly demanding policy actions by all states and by the international community. It would be a great mistake to throw up our hands in despair and say: Copenhagen has been a disappointment, all is lost. In fact all is not yet lost, and there still are technological changes and/or important policies to pursue that can make a difference, that can avert a descent into a Hobbesian world of conflict and destitution. It is such policies and adaptations that provide our ultimate focus.

Daniel D. Perlmutter
University of Pennsylvania

Robert L. Rothstein
Colgate University

Acknowledgments

We would like to thank S. M. (Mike) Miller and Ivan Krakowsky for helpful comments on Chapter 11. The anonymous reviewers for Wiley-Blackwell also provided useful criticism.

Daniel Perlmutter wants to recognize his wife Felice and his children Shira, Saul, and Tova, all of whom deserve grateful acknowledgments for their encouragement and constructive comments at all stages of manuscript preparation. Thanks also to Marshall S. Levine for suggestions and productive conversations.

Robert Rothstein would like to thank John Boulton in London not only for helpful comments on his contributions to this book but also for his encouragement and support throughout. He would also like to thank Marion Lindblad-Goldberg for her very useful comments on Chapters 1 and 11. Finally, he would like to thank Emily Benson of Stakeholder Forum for comments on Chapters 1 and 2 and for the provision of some useful material on the Copenhagen Conference.

Any errors of commission or omission that remain are entirely our responsibility.

Daniel D. Perlmutter and Robert L. Rothstein

1

In the Beginning

The reader will want to know what this book is about, that is how it will deal with the difficult questions about energy and the environment, and specifically with global warming. The plan is to address these matters from two interlocking points of view, considering the technical options that are available, but in the context of the policy decisions and negotiations with which they are intimately linked. In both these arenas, we address an intelligent layperson without requiring any prior expertise on the part of the reader. We outline important policies to pursue that can make a difference and propose specific steps and a list of priorities.

1.1 The Viewpoint Taken

This is a book that seeks to navigate between extremes. We believe that global warming is occurring and that human actions are a major factor in that

The Challenge of Climate Change: Which Way Now? 1st edition.
By Daniel D. Perlmutter and Robert L. Rothstein.

warming, but we are not persuaded that all will be lost if massive policy changes and massive changes in lifestyles are not implemented immediately and everywhere. We have by our reckoning a 10–20-year "window of opportunity" to develop policies that will be effective in facilitating adaptation to existing levels of global warming and mitigating the worst effects of a long-term and very dangerous increase in global temperatures. These policies will be costly, they will require increasingly difficult adaptations as a new energy economy is put in place, and there will be policy "shocks" and costly mistakes along the way, but the changes will not be too costly or too demanding or too surprising *if* we do not use uncertainty and ideological warfare as excuses to procrastinate. We shall discuss all these matters in greater detail in the chapters that follow.

Because many forecasts of climate change have already proved to be too optimistic, that is to underestimate the magnitudes of actual events, it is possible that the window of opportunity could close more rapidly than we now expect. Recognizing the chance that such an unhappy situation could develop, we need to be prepared to implement very rapidly some more extreme and in some cases controversial responses to global warming. For this reason, our policy suggestions in Chapters 9, 10, and 11 follow a two-track strategy: a "normal" track in the next two decades that seeks to establish and sustain a policy process that deepens our ability to deal with and lessen the effects of the warming that has occurred and will continue to occur; and an "abnormal" track, focused initially on enhanced research expenditures, that seeks to avert the worst and to limit the damages from what cannot be averted. There is a loose analogy here with the Obama administration's response to the danger that recently threatened the US financial system as well as those around the globe. In that case massive multidimensional policy responses were called upon to prevent a devastating implosion. A similar emergency response might be needed if the abnormal track becomes a reality, but there are reasonable prospects to avoid such an end.

We want also to be fair in our analysis, not so much to the outright deniers of climate change or to the ideological posturers who seek only partisan advantage from the debate on climate change, but rather to the analysts who believe that we can deal with the problem more effectively and more cheaply by a variety of other means. The science and the politics of climate change are constantly evolving, new information is appearing from a variety of sources, and new configurations of political power and public opinion are constantly emerging. It is also important to emphasize that, whatever the degree of consensus in the scientific and policy communities, the public at large is frequently confused, uncertain, or indifferent to the debate on global warming: they are primarily focused on economic issues (jobs, mortgages, pensions, etc.) and for them the long-range effects of climate change can easily slide to a low position on their political (or personal) agenda. Thus, a recent study showed that global warming was 20th on a list of issues of concern for a typical voter. This has obvious implications for what is or is not likely to be feasible for any administration to contemplate. We shall return to the issue of the effects of public opinion later, especially in Chapter 11.

This may also be a useful point to comment on the December, 2009 controversy concerning the e-mails of a number of scientists at the University of East Anglia that were (illegally) hacked. A few of the e-mails apparently revealed some minor efforts by a few scientists to hide a recent decline in temperatures and to obstruct the publication of some articles by one or another denier.[1] That the hacked e-mails were released just as the Copenhagen Conference opened contributed to public confusion and provided further (if spurious) ammunition for the opponents of rapid movement toward a new energy economy. Thus the Saudi delegate to the Copenhagen Conference cited the released e-mails in his opening speech as a reason to delay action, adding rather bizarrely that in any case the oil exporters should be compensated for any potential income losses if renewable sources began to displace hydrocarbons in the fuel economy. He did not offer parallel compensation by the oil exporters, however, to return their windfall profits to those who have been harmed by them. It is most disappointing that the scientists at East Anglia violated their own professional norms, but in a larger sense it does not make much difference.[2] There is more than abundant physical evidence of global warming, and indeed the World Meteorological Organization responded to the furor by releasing a new report indicating that the decade from 2000 to 2009 was the warmest ever. What is most crucial here, however, is not the failed attempt to refute global warming, but rather the extent to which the debate itself has become an ideological war in which any or all means of resistance seem acceptable to the participants.

These comments are not meant to deny that there are still many uncertainties about the rapidity of warming, about the implications of different degrees of warming, about the best means of dealing with it, and about the costs of different choices. What is *not* uncertain, however, is that global warming has occurred and is occurring at an alarming pace and that it would be irresponsible to delay our responses: there are certainties as well as uncertainties in this debate and the latter should not be used to rationalize inaction. We have tried in what follows to bypass the ideological wars and to focus instead on the need to establish and sustain a policy process that provides insurance – one hopes – against the worst for ourselves and for the most vulnerable. Many of the uncertainties seem unlikely to be resolved for years or decades to come, but they will be more easily managed if a consensus, domestically and internationally, can converge on the need to act *now* against the climate changes that have occurred and are occurring, and to start preparing for those that may occur at an accelerating pace in the years ahead. The obstacles to achieving that consensus are severe, as we shall see, but we shall discuss some of the dangers of failing to do so in what follows.

We have undertaken here to present a practical approach to a broad set of challenges, but recognize also that many readers will want refresher commentary on some of the technical issues, even when they are sufficiently savvy with respect to the political implications. As a result several of the chapters that follow address and explain the scientific ideas that underlie many of the technical proposals that have been put forth. The goal in each case is to supply a background that is sufficient for a fuller understanding of the details involved.

The chapter on *Surveying the Field* puts the discussion of energy into an international framework by addressing matters of supply and demand of the major fossil fuels with respect to time and place. The common units for measuring energy and power (megawatts and gigawatts) and those for measuring petroleum quantities (gallons and barrels) are introduced in order to make possible later quantitative comparisons. They are then used in examining the known facts and doubts on long-term reserves. Above all, the discussion is focused on recognizing that energy is transformable, on the idea that humans do not make energy, but rather that we have learned to transform it from some available form into another one that is more useful for our purposes. This fundamental notion underlies all the impressive engineering that has produced dramatic changes in our social and economic lives in the late nineteenth and twentieth centuries, and is the foundation upon which the later chapters on *Renewable Energy* and *Energy Storage* are built.

In Chapters 5 and 6, solar energy and its connections with the various wind, water power, and biofuel technologies are elucidated, but there are also important questions to be asked and answered as to what energy qualifies as renewable, because a great many political and economic outcomes depend on the definition of this term. Since renewable energy sources are given governmental grants and tax benefits in both domestic and international arenas, it is important to assess whether the beneficiaries of preferential treatment do in fact contribute to the goals that stimulated the legislation in the first place, or whether they are merely hangers-on who seek profit in being identified as purveyors or manufacturers of equipment or goods that can be labeled as renewable.

The chapter on *Energy Storage* surveys the wide range of existing and/or proposed technologies that serve that purpose, highlighting the strengths and weaknesses of each. The reader is led to understand why extensive storage is vital in the development of all the renewable sources, and to recognize that some technologies are actually storage devices, although they may not immediately appear to fall in this category.

Chapter 7 delves into the domestic and international political constraints that have delayed or obstructed a rapid and effective response to climate change. We may know, in general, many of the things we ought to be doing but unless we can remove or reform the political obstacles, we may find ourselves dealing with global warming as an emerging catastrophe. Chapter 7 is both critical about the negotiating failures of the past and prescriptive about what might be done to improve the prospects for more successful future negotiations. The final chapter, *Prospects after Copenhagen*, adds some further comments on the possibilities of reform.

Chapters 8, 9, and 10 are oriented to the near and not so near future: we propose a series of steps that ought to be taken and put them in the form of a list of priorities with dates attached. Many of these moves are in and into the technology realm, but they also include steps of negotiation and policy change internationally as well as domestically. It appears at times that negotiating with our own US Congress is as difficult, sometimes more difficult, than finding agreement abroad.

Chapter 11 discusses what the Copenhagen Conference of December, 2009 did and did not accomplish. Was it a useful exercise in international policy-making or a waste of time and resources? We shall also return there to some of the political questions left unresolved in Chapter 7, especially about the effects of public opinion and the possibilities of reforms that might generate more effective responses to climate change.

The central theme that is to be developed may be summarized as "It's not too late – yet". There is a window of opportunity that covers a decade or two, during which a variety of technical and political moves can offer various degrees of prevention, amelioration, and/or remedy for the most egregious harm to our planet and its inhabitants. Our approach is pragmatic, suggesting priorities for feasible alternatives, emphasizing decisions that are moral as well as practical. We will attempt to show that fairness and intergenerational equity are central and necessary components of any decisions that hope to bring about change.

1.2 What is Your Problem?

There is a broad consensus in both the scientific and policy communities that something needs to be done – and soon – about the world-wide energy utilization that is bringing about continuing global changes leading to serious environmental deterioration. An uncontrolled global warming would create catastrophic consequences for the earth and its various political, economic, and social subsystems. There is, however, widespread disagreement about what should be done when, and how the costs of any policy choices should be distributed. Disagreement is hardly surprising given continuing scientific uncertainties as to the extent of anticipated changes, the vast time scales involved, and the potentially enormous costs of making a transition to a new energy economy and perhaps even to a new or significantly altered theory of economic growth. The uncertainties are compounded because the necessary changes are likely to require an unprecedented degree of international cooperation and a degree of domestic policy consensus that may be unlikely as powerful interest groups resist changes to the old order. In addition, one needs to recognize the obvious fact that it is normal economic activity – growth, development, trade, industrialization – that is at the root of the energy and environmental crises, further complicating the problem of achieving consensual change.

The potentially devastating consequences of doing nothing about energy and the environment are well documented and frequently cited. Less recognition has been given to the positive strides that have been made in this regard in the last three decades, emerging signs of change that are significant and notable. Major contributions have come forth, for example, following the organization in 1974 of the Worldwatch Institute, whose mission statement includes:

> The Worldwatch Institute is an independent research organization that works toward the evolution of an environmentally sustainable and socially just society, in which the needs of all people are met without threatening the health of the natural environment or the well-being of future generations. Through accessible, fact-based analysis of critical global issues, Worldwatch helps to inform people around the world about the complex interactions between people, nature, and economies. Worldwatch focuses on the underlying causes of and practical solutions to the world's problems, in order to inspire people to demand new policies, investment patterns, and lifestyle choices.

Various publications of the Worldwatch Institute[3] have been warning about environmental degradation for decades. Disappearing forests, eroding soils, collapsing fisheries, water shortages, melting glaciers, the disappearance of plant and animal species, and increased global warming are all threats arising from a multidimensional failure to act.

These gross changes must sooner or later produce major societal responses, some of which will have arisen from more than a single cause. Classifying a given response as economic or political in origin is arbitrary, since the "spillover" from one to any or all of the others can be significant. And yet, in spite of the well-publicized warnings, the reactions have generally been mild and largely unresponsive. There are a variety of reasons for this: uncertainties about what choices to make, scientific and ideological disputes, fears about excessive costs especially in a very difficult and dangerous economic crisis, the ease with which "resistance fronts" can delay or thwart action both domestically and internationally, and of course the enormously high stakes involved in choosing badly or not choosing at all.

The lack of any strong actions on the part of governments has created a vacuum into which political activists have moved. Former Vice President Al Gore, for example, has made a notable effort to raise public support for immediate and massive responses to global warming and the associated components of environmental decline. Gore's warnings have been published in a well-known book[4] and in an associated documentary that won an academy award. His contributions were recognized in 2007 by the Nobel Prize.

By shifting his life's work from one political arena to another, Gore attracted world-wide attention to a movement that has at least begun to shift attitudes among the general public. Hardly a day goes by that newspapers and television reports do not feature a story about some aspect of environmental concern or renewable energy. Advertisements by major corporations now feature claims about their products or policy developments intended to help this cause. It appears that there has been in the last decade at least the beginning of an attitude shift concerning the need to do something about these threats. It must be added, however, that strident warnings have been met by mixed reactions in some circles. Depicting threats in a manner that seems overwhelming and unstoppable without massive policy changes, the results of which are inherently uncertain, may elicit an abreaction, a deeper pessimism about the value of doing anything at all, reinforcing in effect the cognitive conservatism of most individuals who are reluctant to challenge the conventional wisdom or to alter basic views.[5]

As will be shown in detail in the coming chapters, our continued dependence on energy from hydrocarbon fuels – oil, gas, and coal – is the major cause of global warming as levels of carbon dioxide in the atmosphere continue to grow. Though this effect is generally understood, efforts to achieve a viable international agreement to cut emissions have foundered. The Obama administration's commitment to become involved in the negotiations to strengthen the Kyoto agreement "in a robust way," which initially generated a surge of optimism, seems in practice to have retreated somewhat from campaign rhetoric – not surprising, given the pressure of other issues and the complexity of creating a domestic consensus.[6] The issues are, of course, also intimately tied to political and economic considerations: higher oil prices, increasing dependence on potentially unreliable suppliers in potentially unstable areas of the world, increasing economic nationalism and resource nationalism, competition not only between producers and consumers but also between different consumers, increasing inequality and a growing sense of unfairness, and the likelihood of increased conflict within and between states fighting to control resources as paying for oil leaves less money to deal with other pressing domestic needs.

There are other political pressures worth noting, apart from the ability of powerful interest groups to thwart change and apart from the fact that virtually all political systems, especially democratic ones, have great difficulty in taking seriously any long-range plans. To these must be added the current economic crisis, whose effects may be with us for a decade. Further, there are clear temptations to hold closely any technological breakthroughs that do occur, considering them as valued intellectual property to be sold only at a very expensive price. Rather than treat the breakthrough as a public good to be shared, the creator may seek to hoard knowledge and earn the monopoly rents that may become available. Such behavior is reminiscent of the so-called "beggar thy neighbor" policies of the 1930s, which sought to export problems rather than resolve them cooperatively, reinforced both then and now by the absence of an international institution to facilitate such cooperation, as well as by the absence of strong leadership from the rich and powerful.

An additional unexpected result might be described as the perverse effect of some seemingly good development. If the demand for oil decreases and its price falls, for example, the incentive to invest in alternative energy is reduced and may lead to a decline in government subsidies that would make it even harder for the alternatives to compete and survive. In effect, policy-making that is a captive of events – this year's recession, next year's boom – is episodic and inconsistent policymaking. There is also the possible effect of another "moral hazard": excessive enthusiasm for this year's quick fix, say carbon capture, may delay or undermine other painful policies that might begin to reduce emissions more quickly.

Given that the products of combustion of hydrocarbons are the major contributors to global warming – which is increasingly the consensus among a large majority of reputable scientists – and if the result is the increasing likelihood that many of the dangers and threats noted above will create global, regional, and national instability, declining standards of living, and (perhaps) a permanent state of crisis, fear, and anxiety, why has it been so difficult to forge common policies to avert such threats? It is not difficult to lay out what a sensible policy package should look like for the US or other nations or indeed for the international community as a whole, but putting those policies into practice has been and may continue to be an entirely different matter. We shall explore some of the reasons for this in the next section.

1.3 The Challenges We Face

We have already alluded to some of the obstacles that impede the development of effective national and international policymaking, to wit:

1. Powerful and rich interest groups have been and still are an obstacle to major reform of the existing energy economy.
2. There are strong divergences of interests between the developed countries and those of the third world, who distrust any notion of moving

away from a development strategy based on rapid industrialization and intensive use of conventional sources of energy.

3 The potential costs are vast and unappealing in a period of economic turmoil and fear.

4 Public opinion is not (yet?) a powerful voice for action and sacrifice.

5 No political system is good at taking long-run needs seriously, that is, paying heavily now to ensure future benefits or to avoid dangers that may not ever develop. The easier option is to "muddle through," to hope that something will turn up, or that someone else will bear the brunt of whatever transpires.

These are perhaps the most frequently cited obstacles to achieving a consensus, both domestically and internationally, on the need to implement – not merely assert the need for – a comprehensive strategy to deal with linked environmental and energy crises. But clearly one needs to try to deepen and extend this analysis beyond the mere recitation of familiar obstacles. There are avenues for persuading governments and their publics of the need to pay now or to sacrifice now in order to achieve uncertain benefits in the future, even the distant future. Summers has noted, for example, that the costs and benefits of a policy in the future are usually given less weight because of the tendency to value benefits today more than tomorrow, and because we believe that we are likely to get $1 worth of goods in the future by spending less than $1 today.[7] He points to a compensating factor however, the moral obligation "most of us have about our obligation to posterity." It should also be pointed out to the reluctant long-term investor that we have no guarantee of greater riches in the future, especially if our fears of global damage are in fact realized because of our neglect of the options available today.

Intergenerational equity has always rested on a kind of tacit norm about distributive justice: each generation accepts informal obligations toward the future because of its own expectations of future reciprocity. In the US this idea has been the basis of the programs for Social Security and Medicare, among others. In an environmental sense, this has implied that each generation can make fair use of land and resources for its own needs but cannot or should not injure future users by unnecessarily undermining or degrading the long-term productivity of that land or those resources. This was captured in the Brundtland Commission's definition[8] of "sustainable development" as development "that meets the needs of the present without compromising the ability of future generations to meet their own needs." This standard underlies our commitment to preservation of park lands and protected wilderness areas, national policies that have long had broad acceptance and approval in our societies.

These formulations can serve us as guidelines or standards, but individual applications must still address operational questions in practice. How much should we pay now for uncertain future benefits? Who decides how and when to make what investments? Who should bear most of the costs now and enjoy most of the benefits in the future? How "future" is the future, that is, are we thinking 10 years ahead, or a generation – or what? There seems no alternative, at least in democratic societies, to accepting the fact that the decisions will be the uncertain outcomes of the political process, a process in which the needs of future generations will only be one interest among many.

One of the least discussed aspects of the current economic crisis is the level of anxiety and fear that it has generated among citizens at all levels of society, not merely the usual victims, the poor and powerless, but also the (formerly?) rich and confident. Fears about losing jobs, incomes, retirement benefits, a whole standard of living, are pervasive and not irrational. Unfortunately, playing on this fear and anxiety and attempting to manipulate these emotions for political or economic gain (especially through the popular media), exacerbates the problem. In this context, it is essential to cultivate in the public and among government leaders a willingness to think about future obligations and to counter the familiar biases associated with decisionmaking under uncertainty[9] by emphasizing repeatedly that there are options to be considered. The point here is not merely that arguing for inter-generational equity might remove a layer of difficulty and complexity in the negotiating process, but rather that its use could soften in some layers of the population (older, wealthier, influential) an otherwise consuming focus on their personal economic welfare.

It will not suffice to offer a few platitudes about the long-run consequences of excessive selfishness or reciprocal nationalisms; one also needs practical suggestions about how to create a viable negotiating process in such circumstances. We shall discuss this issue in more detail in a subsequent chapter. Here it is sufficient to say that "big bang" negotiations in a global conference setting may be a negotiating bridge too far, guaranteeing only stalemate or pious commitments that are rarely implemented. "Normal" incrementalism is also likely to be inadequate, which implies the need for a deeper, more sustained kind of incrementalism with consensual long-run goals approached by regular steps in a jointly agreed direction.

Another obstacle to the creation of viable international agreement, and one whose significance has not always been understood, is the absence of widely shared knowledge about the causes of a problem and what needs to be done to resolve it. Haas[10] has described this as "consensual knowledge," which is socially constructed, subject to testing and evaluation, and thus different from ideology in facing constant challenges. It has been difficult to establish such knowledge in the present context because of factual complexity and the extended time periods involved. Superimposed on these factors is the occasional dissent from the majority scientific opinions that can be used as reasons or rationalizations to resist action. Note, for example, the statement by three scholars that "the only reliable knowledge is that current understandings of the problem will be obsolete in ten or twenty years."[11] The tendency to pick and choose among different pieces of evidence and to ignore conflicting evidence is also an impediment to agreement. So too is the desire of politicians for certainties in an environment dominated by probabilities and uncertainties.

That knowledge is ambiguous, uncertain, and contested is not, however, an argument for doing nothing. It is, rather, an argument for an even stronger effort to create consensual knowledge especially by creating large panels of widely recognized experts to seek and publicize the best available knowledge and to provide enhanced funding for serious research projects. The success of the Intergovernmental Panel on Climate Change (IPCC), which won a Nobel Prize a few years ago, is illustrative of what can and should be done: its papers were far more authoritative and balanced than contrary efforts either by environmental alarmists or environmental skeptics and were at least helpful in supporting governmental decisions to take relatively strong policy positions on the issues. Of course, it would be easier to achieve consensus about strong commitments to action if one had something like the moon landing project in the US to rally support and engender enthusiasm. That project had reasonably strong consensual knowledge about how to pursue the task, strong leadership and public support, and the vision of a "race" with a determined enemy to justify the effort. Unfortunately, no such conjunction of circumstances exists either domestically or internationally to galvanize a similar outcome with regard to energy or the environment.[12] Some suggest that only an environmental disaster in the developed world would suffice, something akin to the Asian tsunami but hitting say London, or Paris, or Miami. That is, of course, an extraordinarily costly and irresponsible way to engender effective policy responses. Besides, the enormous surge in aid giving immediately after the tsunami does not appear to have had a lasting effect on either regional governments or international institutions. With few exceptions (mostly in regard to installing better technical sensors) most responders seem to have retreated into business as usual.

Consensus begins at home; that is, the basic obstacle to achieving agreement, which we have already noted at various points, is the play of domestic politics. In effect, the problems we are discussing are sub-system dominant: international organizations, non-governmental organizations (NGO), and any other international entities lack the power and the resources to do much by themselves.[13] There are no clever tactics or stratagems to overcome this problem, but it is worth emphasizing an obvious point: with or without domestic consensus on the issues, strong and determined leadership is the crucial variable that permits a degree of hope in the current situation. Thus the replacement of the resistant Bush–Cheney administration by the Obama–Biden administration *may* allow the US to reassert leadership in the energy and environmental arenas and, finally, to receive a fair and respectful hearing for its policies and positions. But determined action may have to wait upon successful navigation through the current economic crisis.

The environment has rarely been a "front burner" issue for most governments and until recently even the energy issue has only been front and center when there has been a sudden surge in oil prices. As a result, conferences on these issues have tended to be dominated by environmental advocates, industry advocates, environmental scientists, and mid-level government bureaucrats. Thus, for example, in retrospect many officials who participated in negotiating the Kyoto Protocol in 1997 "... now say they see it as weak and naïve about political and economic realities." As a British official said, "In Kyoto, we made a lot of promises to each other, but we hadn't done the domestic politics and that is why Kyoto ... has ultimately been so fragile."[14] There are two simple points here: negotiations have to be driven by committed and high-level leadership, especially from the US, and expertise about domestic political constraints is crucial if one wants agreements that will be ratified and implemented. In sum, the obstacles are severe and the tools to deal with them are as yet underdeveloped.

But to acknowledge the problems that face us does not condemn us to surrender to them, and it is the purpose of this book to offer constructive approaches that can mitigate unwelcome effects and even avert them where possible. Our suggestions will focus on both policy and science technology, whichever choice or combination of choices will serve these ends.

Notes and References

1. There is an excellent column about this incident by Thomas L. Friedman, see "Going Cheney on Climate," *New York Times*, December 9, 2009, p. A37.
2. This is not meant to deny that the furor over the leaked e-mails *did* make some difference. Thus, in an environment where public attention is focused on the economy and the skeptics and deniers speak with an inappropriate degree of certainty, a poll in the months after the incident indicated that the share of the public who do not believe climate change is happening increased from 15% to 25% and the percentage who do think it is happening and man-made dropped from 41% to 25%. See "Greener than Thou," *The Economist*, February 13, 2010, p. 61.
3. The diametrically conflicting views of the Worldwatch Institute and the Competitive Enterprise Institute are discussed in William D. Sunderlin, *Ideology, Social Theory, and the Environment* (Lanham, MD: Rowman and Littlefield, Inc., 2004), pp. 178–9.
4. Al Gore, *An Inconvenient Truth* (Emmaus, PA: Rodale, 2006).
5. Andrew C. Revkin, "In Debate on Climate Change, Exaggeration is a Common Pitfall," *New York Times*, February 25, 2009, p. A14.
6. See Elisabeth Rosenthal, "At U.N. Talks on Climate, Plans by the U.S. Raise Qualms," *New York Times*, April 9, 2009, p. A14.
7. Lawrence Summers, "Foreword," in Joseph E. Aldy and Robert N. Stavins, eds, *Architectures for Agreement: Addressing Global Climate Change in the Post-Kyoto World* (Cambridge, UK: Cambridge University Press, 2007), p. xix.
8. Brundtland Commission, Report of the World Commission on Environment and Development, published as *Our Common Future* (Oxford, UK: Oxford University Press, 1987).
9. For comments on the problems of decisionmaking under uncertainty and in an environment of fearfulness, see Daniel Gardner, *The Science of Fear* (New York: Dutton, 2008), p. 39ff.
10. Ernst B. Haas, *When Knowledge is Power: Three Models of Change in International Organizations* (Berkeley, CA: University of California Press, 1990), p. 21.
11. Peter M. Haas, Robert O. Keohane, and Mark A. Levy, "Improving the Effectiveness of International Environmental Institutions," in Peter M. Haas, Robert O. Keohane, and Marc A. Levy, eds, *Institutions for the Earth* (Cambridge, MA: MIT Press, 1993), p. 410.
12. In any case, perhaps the moon landing is an inappropriate analogy because we do not have a single aim in the policy debates on climate change but rather many aims, some of which compete with each other.

13. Peter M. Haas, Robert O. Keohane, and Mark A. Levy, "Improving the Effectiveness of International Environmental Institutions," op. cit., pp. 397–426.
14. Both quotes are from Elisabeth Rosenthal, "Obama's Backing Increases Hopes for Climate Pact," *New York Times*, March 1, 2009, p. 10.

2

A View of Geopolitics

There is a massive and growing literature on global warming and from almost every imaginable perspective. In the circumstances, one might ask what can be learned from re-focusing the current debate in terms of an earlier controversy about energy and the environment: the limits to growth debate of the 1970s. There are in fact some useful lessons to be gleaned from that earlier debate, and they will be brought out in what follows in this chapter. Moreover, there are benefits in viewing the current discussion from the perspective of international *public goods*, a category that has heretofore barely existed. In order to provide a new kind of analytical background for discussion, we will also explore what this framework can teach us about the need to create and sustain such public goods. These two approaches to the topics of climate change and control form the basic structure of this chapter, augmented in a concluding section on the policy debates that prepares the context for the more technical chapters that follow.

The Challenge of Climate Change: Which Way Now? 1st edition.
By Daniel D. Perlmutter and Robert L. Rothstein.

2.1 Are There Limits to Growth?

The Limits to Growth was published in 1972 and was an instantaneous *cause célèbre*.[1] Using a variety of sophisticated computer modeling techniques, the authors argued that the continuation of simultaneous trends in population growth, industrialization, food consumption, resource use, and pollution would lead to an inevitable collapse of the world economy and global society sometime in the next 100 years. Subsequent publications 20 and then 30 years later slightly modified some of the more dire forecasts but the basic message remained the same: almost without thinking about it, the earth, or more particularly developed industrial societies, had overshot the limits at which renewable resources can renew themselves and the rate at which society can change from reliance on non-renewable to renewable resources.[2] It was argued that crucial resources like food, water, and oil would disappear, at varying rates depending on different assumptions about the rapidity of growth, and rising levels of pollution would become unsustainable. Chaos and collapse could only be averted by slowing the exponential growth of population and the rapid development of industrialization.

A new model of growth was seemingly necessary, although it was not clear what that model might be or whether any of the alternative models were viable or more acceptable, especially to the Third World, than more of the same. The oil crisis that erupted after the 1973 "Yom Kippur" war in the Middle East – a 500 percent increase in oil prices within a few short years – undoubtedly contributed to the sense of gloom and doom pervading the debate about resource depletion, rising prices, and potential global conflict and instability. Moreover, the oil crisis of the time elicited some extreme reactions on the part of consuming countries, which are a useful warning about the dangers of hysteria under pressure: some advocated the invasion of Saudi Arabia and the other Gulf countries on the principle that they did not own the resources under their land; others advocated using food as a weapon, apparently to starve the Saudis and the others into cutting prices or doing so out of fear of what the US or others might do. These recipes for disaster and chaos never apparently got much beyond rhetoric in magazines like *Commentary* but they did scare the Saudis into trying to grow their own feed grains at very high cost in very arid deserts.

The Limits to Growth did make a significant contribution to the emergence and deepening of the environmental movement and it apparently provided a warning about the implications of the prevailing model of economic growth.

The warning was not that everything was suddenly going to run out or that collapse was inevitable on a date certain. Rather, the warning was in the implicit form of speed limits: if the authors were correct, we were exceeding safe speeds in our use of resources and our pressures on the environment and we, all of us, would pay a severe price if we did not mend our ways quickly. But *The Limits to Growth* and many other models (or polemics) of doom that it generated and seemed to justify also elicited strongly negative reactions from several major sources: the Third World, many neo-classical economists, and some dissident scientists who argued – standing the doomsters' arguments virtually on their head – that affluence and industrialization were ultimately good for the environment and for the poor.

The Third World position was largely ideological; it is difficult to find any clearly articulated, conceptually sophisticated analyses of the implications of real or ostensible limits to growth by Third World analysts. What many Third World leaders and intellectuals seemed to think, almost reflexively, was that the whole debate was a conspiracy on the part of the rich developed countries to maintain their power and their prosperity by telling the developing countries that they had to give up a model of growth that had worked well in the past and seemed to be the only model that would put the developing countries on the same road to wealth and rising standards of income. If they gave up that model, one component of which was rapid industrialization, what could replace it? If they used resources more sparingly how would they overcome poverty?

In effect, the Third World agenda did not focus on ozone depletion or global warming (except by some threatened island states) but rather on such issues as declining agricultural productivity, toxic chemical contamination, and, above all, economic growth, job creation, preferential market access, and more foreign aid. In addition, there was a sustained effort to attribute blame for the energy and environmental crises to the developed world and to insist that the developed countries were thus responsible for the full costs of repairing the damage. That China, India, and other developing countries were becoming major contributors to both crises was considered a future problem, not the central issue of who was responsible for creating the problems that now existed.

Were there alternative growth strategies available that dealt equitably and effectively with both sets of concerns? Certainly the no-growth models postulated by Herman Daly were non-starters.[3] Daly argued that growth was not always good and that the idea of "sustainable development", which had become a watchword among environmentalists, especially in the United Nations (UN) system, was a meaningless phrase, but he left unanswered questions that were directly relevant for poor countries with limited human capital or financial resources. Other models, such as one that emphasized fulfilling basic human needs before a heavy focus on industrialization, were greeted with suspicion not only by Third World elites but also by conventional neo-classical economists. Hostility to such a strategy, based in part on the unwarranted assumption that it was anti-growth, first appeared in the 1970s as a response to the failure of standard growth theory to reduce poverty significantly in the Third World.

Some part of the Third World resistance was simply the desire to compel the developed countries to increase foreign aid and direct investment, augmented by fears among their elites that any change might cause them to lose their direct control over how and where aid monies were to be spent. Another part of the resistance reflected the sense that concern for the environment was a luxury that they could not yet afford. This rather paralleled the once fashionable argument by many governing elites that democracy and political competition were luxuries that poor countries could not afford. This self-interested argument might have made more sense if the governing elites actually used their power for the common good, but unfortunately much of their power was misused merely to retain power and reward friends and potential enemies. Moreover, given the fact that many of the resources that were going to be in short supply were located in the Third World, there was also the hope that they might in fact benefit from the resulting price surges, as had happened with oil.

The main critics of the limits to growth argument were conventional neo-classical economists such as Julian Simon. Their argument was in essence quite simple: warnings about any limits to growth were oversimplified because they focused on too few variables and failed to take into account how human technological ingenuity and the normal operation of the market would respond to increasing scarcity. Human creativity was, according to Simon, "the ultimate resource" and is essentially inexhaustible. Thus, he argued, *The Limits to Growth* was "a fascinating example of how scientific work can be outrageously bad and yet be very influential."[4] Moreover, he argued that the price of raw materials and energy has fallen steadily in real terms (short-run "blips" aside), food supplies were steadily growing even in most parts of the Third World, more land was being brought into use, pollution could not be much of a danger if life expectancies continued to improve, and population growth was a net benefit to the developed countries and to the developing countries if kept moderate. His position was supported by Lomborg,[5] who also focused on the statistics showing less world poverty and longer life spans. In sum, the environmental "doomsters" were "killjoys riding bandwagons to political power."[6] One other aspect of the anti-doomsters' argument was that adaptation in some areas (for example, conservation and efficiency improvements) was faster than doomsters seemed to anticipate, implying that there was more time and space to innovate and adjust, especially in areas like the more widespread use of fertilizers, irrigation, and new medicines.

These arguments provided a useful critique of some of the weaknesses of the limits to growth models: oversimplification was an especially apt charge because too much was left out or ignored, politically, socially, and economically. Still, while the market has produced great prosperity, the critics who rested their case on how it should or might work were also leaving out some key considerations. Thus markets may respond to price signals but they frequently do so with a serious time lag; political systems and public opinion may be even more laggardly, especially as strong interests resist adaptation and intellectual uncertainty still prevails. We cannot simply take it on faith that the human capacity to innovate is inexhaustible and will provide solutions in time, the possible inequity of some adaptations may also be another source of delay, and crises may occur with sudden, sharp shocks, not gradually with time for adjustment.

One problem with this debate, especially in its earlier years, is that it had a strong tendency to degenerate into ideological posturing, the free marketers deriding the lack of economic sophistication of the doomsters and the environmentalists seeing only the need to alter current policies radically to avoid disaster. It was a case of two ships passing in the night, of competing paradigms each barely acknowledging the existence or importance of the other.[7] Fortunately, the debate has become less ideological as evidence of the accumulation of carbon dioxide in the global environment has become more widely accepted, largely legitimized by the periodic reports of the prestigious Intergovernmental Panel on Climate Change. As a result many proponents on both sides have recognized that there is a price to pay for emissions, and some, seeing the dangers, morally and politically, of ignoring the issue of equity for the poor everywhere, are prepared to take into account some of the concerns of the developing countries.

This is not to say that there is complete consensus on all these issues. To the contrary, there is some degree of dissent on virtually every issue, but the differences are not as stark as they once were. Even the most avid free marketers now see the need for action on the environment, and most of the environmentalists have come to realize that there is a need for some kind of income growth. Many of the disputes now concern specific policy choices and the difficult trade-offs between different values, even as both sides recognize that the still unresolved scientific and technical issues that need to be clarified should not be used as a rationalization for complete inaction. The overarching issue is how to reconcile two equally valid but potentially divergent goals: continued income growth and the alleviation of poverty versus protection of the environment against the worst effects of growth and the continued dependence on carbon-based fuels. There is some hope that a new kind of qualitative growth strategy can be devised, but even if it is devised, it is far from clear that it will be accepted by either developed or developing countries in the light of perilous economic circumstances and the resistance of established interests.

There are and have always been scientists opposed to the simple, perhaps too simple, proposition that population growth and affluence supported by technological development would lead inexorably to environmental disaster. Jesse Ausubel and Paul Waggoner, respected environmental scientists, are in the forefront of the dissident voices. They have argued that declining consumption of both energy (especially carbon-based fuels) and goods per unit of gross domestic product (GDP) offers hope that economic development and improvements of environmental quality can occur at the same time.[8] The diminishing carbon intensity of economic activity, largely because of efficiency gains, has characterized most countries in the period 1980 to 2006. As a result, "all of the analyses imply that over the next 100 years the human economy will squeeze most of the carbon out of its system and move via natural gas to a hydrogen economy."[9] In short, both population and affluence can continue to grow without a proportionately greater environmental impact.

Ausubel has also argued, as have spokespersons for many of the oil companies, that our current energy system, based on carbon fuels, is likely to continue well into the future. He says "Energy systems evolve with a particular logic, gradually, and they don't suddenly morph into something different."[10] As Tierney says, this implies that the absence of a green revolution in energy is good because "The richer everyone gets, the greener the planet will be in the long run."[11] The argument is that "as incomes go up, people often focus first on cleaning up their drinking water, and then later on air pollutants like sulfur dioxide." Thus, presumably, nothing should be done to impede rapid industrialization, because the immediately bad environmental consequences of this will soon be reversed.

What can be said about this argument? In the first place, taken in context with the previous arguments, it is not difficult to see why even reasonable and intelligent observers can disagree about global warming and the energy crisis. There are apparently reasonable arguments in support of many different policies and what gets chosen in many cases is likely to reflect personal ideologies, interpretations of self-interest, and even which set of statistics to emphasize. Secondly, in many ways the Ausubel and Waggoner argument offers the developing countries a dream come true: not only are they told that the standard approach to economic development is still the best way to growth and prosperity – no need to gamble on untried new theories – but also that it will someday generate better environmental outcomes. Finally, even if intensities of use steadily decline as Ausubel and Waggoner argue, what will happen to global warming while this acceleration of industrialization goes on? What if this faith that the market will provide adequate substitutes for increasingly scarce and expensive resources proves excessively optimistic? And what if economic turmoil, political conflict, and social instability derail the trend toward "decarbonization" before it produces affluence and environmental improvements? In short, while providing us with a picture of long-run success, the Ausubel–Waggoner position ignores all the short-run dangers and difficulties that might render their conclusions merely academic.

There seem to be two simple lessons from this brief discussion of the limits to growth debate. The first is that the kind of alarmist predictions and cataclysmic forecasts that some environmentalists indulged in on the basis of *The Limits to Growth* book, a kind of "worst case" analysis, actually had the opposite of the intended effect. As an illustration, note the comment by Byron: "Civilization's foundation is fatally insecure and its collapse is imminent."[12] If our circumstances were perceived to be so bad with so many vast and costly changes that had to be made so quickly, even before the science was perceived as conclusive, it is not surprising that many people would shrug the message off and go back to business as usual. A more carefully devised argument with a clear set of priorities and a clear sense of the likely trade-offs might have produced a better result sooner. In the same sense, arguments that tell us all is well doing "what comes naturally" may also deflect our attention away from what needs to be done now. The second lesson bolsters the first. The ideological, political, and economic warfare between two divergent paradigms delayed an effective policy response: short-run price signals, such as the declining or flat price of oil in much of the 1980s and 1990s, were used as an excuse to dismiss the warnings of the dangers ahead. The long list of obvious things that should have been done after the experiences of the price shocks of the 1970s – conservation, increased efficiency, higher mileage standards for automobiles, enhanced spending for research and development of alternative energy sources – were ignored because it was easier and less costly to do nothing and to dismiss warnings that were initially oversimplified and overstated but nonetheless certainly worth much more serious consideration. Debate is useful if it clarifies, not if it obscures the need for action.

2.2 Public Goods and Public "Bads"

When we speak of public goods we mean goods that are non-rivals, that is consumption by one party does not reduce the supply of the goods available to others. Public goods are also non-excludable, that is no one can be excluded from using the good. Conventionally, public goods include such things as national defense, lighthouses, the public school system, and the interstate highway system. There are, however, other variants of public goods. Club goods are non-rival but excludable: examples would include, say, national parks, public golf courses, or toll roads that charge a user's fee. Another variant that is rivalrous but non-excludable is described as a *common pool* resource: examples would include deep sea fisheries or the exhaustion of shared grazing areas by overuse. This latter event has often been referred to as the "tragedy of the commons." Because markets are uncoordinated and driven by individual self-interest, they sometimes fail to produce or underproduce public goods. As a result, public goods have usually been produced by the state but they can also sometimes be produced by non-state collective actors such as churches or non-governmental organizations like the Sierra Club or the Red Cross – or of course not produced at all. Actors who make use of a good or benefit from it but do not pay for its creation or maintenance are referred to as *free riders*. The incentive to defect and not pay a fair share of the cost of producing the public good can be very high, if the benefits are not great or sanctions for non-compliance are weak or easy to evade.

There are obvious questions that arise in regard to the production of public goods. How much of a good needs to be produced? How is its production to be financed? Given uncertainties about the future scientific–technological knowledge base, there is also a question about how to acquire what kind of knowledge to justify the creation of a public good that will be effectively implemented. In fact, uncertainties about costs and the risks of non-compliance or unanticipated developments can undermine production of public goods. Why pay such costs or commit to compliance if the risks and costs are high and the likelihood of non-compliance is great?

The discussion of public goods has usually been limited to national states and indeed emerged in the discipline of public finance. But what we have seen increasingly in the past 30 or 40 years with globalization are that negative externalities, the extra costs of economic actions, are increasingly borne by neighbors across borders. Thus the quest for global public goods has become a new focus of interest not only because of the "overspill" of externalities across borders but also because global systemic risks have become an increasing area of concern. The current economic crisis and the obvious inadequacy of the global financial architecture to deal with it illustrate the danger in one area of the under-provision of such goods. The increasing weaknesses of the regime to deter nuclear proliferation is another illustration of the consequences of failing to devise an effective international public good. The problem in the international system is, of course, that free riding can become pervasive, cheating is too easy without a strong sanctions regime, and there are no authoritative bodies to deter violations or to impose penalties for non-compliance.

The gap between what is needed and what can be provided is especially wide in international affairs, but not necessarily completely unbridgeable. There have indeed been a few successes in achieving international public goods. As Barrett has ably demonstrated, the Montreal Protocol on ozone depletion was a success because of a variety of factors: it put limits on *all* states, not just the developed countries; the cap on emissions was permanent, not temporary or transitional; and there was a strong deterrent to free riding in that the compliance mechanism was quite strong, with threats of a suspension of aid to developing countries or trade sanctions. In contrast, the Kyoto agreement on climate change had none of these conditions, the US never signed the agreement, and it was badly implemented because the compliance regime was weak. Unfortunately, it has been difficult to replicate the conditions that made Montreal a rare negotiating success. The international agreement on Antarctica has also been reasonably successful, perhaps because provisions about the duration of the treaty and its renegotiation provided the participants with a necessary degree of flexibility.[13] Insofar as the parties to an agreement have a shared history of cooperation or are bound to continue to interact in important ways in the future, a concern for maintaining a good reputation about fulfilling commitments, and establishing a degree of implementation credibility, may also help some agreements to survive and prosper, even when one or both sides do not see a strong national interest in doing so. It is also very important that all parties to an agreement see it as fair and equitable, although agreeing on what these terms mean may be more difficult than reaching agreement in the first place.

We have had a scientific and technical debate on the extent of the problems in the energy and environmental arena, a political debate over the means to achieve cooperation and the distribution of costs and benefits, and a moral debate about equity, responsibilities to future generations, and the intrinsic claims of the environment itself. There are disagreements about all these issues, some severe but others where there has been a degree of convergence. The one thing about which virtually all agree is that none of the problems in these complex realms can be resolved unilaterally or by a relatively small group of states: what is and will be necessary is a very strong and virtually unprecedented degree of international cooperation. Global or regional crises cannot be localized or contained within national borders, as the problems of acid rain and Chernobyl illustrate in different ways. And even if collective security has proved unworkable or ineffective, some version of collective environmental and energy security may have to be devised.

We have already noted most of the domestic obstacles to achieving desired or desirable levels of cooperation. Here we focus on the additional obstacles to international cooperation, which alone or mixed together make progress difficult even among well-intentioned participants; specifically:

1. The absence of intellectual consensus.
2. A focus on the short run.
3. The desire to protect an ever-eroding degree of sovereignty.
4. Weak and poor international institutions that cannot easily impose sanctions for deviant behavior.
5. Different ideologies and interpretations of national and international interests.
6. Power asymmetries.
7. Local and regional conflicts.
8. Factual uncertainties that raise the risks of making clear policy choices.
9. The constraints and anxieties created by the current economic crisis.

Moreover, defectors and free riders are an inevitable danger in such circumstances. In short, we have a very large gap between the demand for more effective means to achieve necessary and beneficial degrees of cooperation and the capacity to supply it internationally. As noted earlier, we are dealing with essentially value-laden conflicts that can only be resolved through the political process. We can perhaps gain a better perspective on these issues if we turn to a brief discussion of a very similar analytical problem – the insufficient provision of international public goods. Or perhaps we will learn something even more useful: that a different kind of approach to achieving cooperation may be necessary.

For the most part international public goods have been much harder to produce than domestic public goods. In addition to the obstacles already noted, a few other hindrances are worth a brief comment. Inter-generational equity is of course a weaker force between different countries than it is in any one nation. There are certainly some shared norms and traditions in the international system but in the present context they are as yet rudimentary and mostly limited to a relatively small group of committed environmentalists. Exhortations or dire warnings of imminent catastrophe will not suffice when there are too many more immediate problems that require attention. Finally, there is the basic dilemma at the root of international relations: the absence of trust is much greater than it is in domestic systems, the possibility that the other or others will renege on commitments is greater, it is more difficult to know or discover what other parties are up to, interests vary greatly on many dimensions, and the initial conditions that led to negotiations may alter in fundamental ways, generating the need or the desire to grasp new opportunities or ward off new threats. One result of this is that the focus sometimes shifts from the creation of hard-to-negotiate public goods to the negotiation of agreements to ward off consensually agreed-upon public "bads." However, this too has not been easy: note, for example, the difficulty of achieving a critical level of consensus about stopping the proliferation of nuclear weapons, as with Iran and North Korea.

Some scholars have argued that international institutions can step in at this point to help negotiate and sustain an international agreement. According to their arguments, such institutions can help to overcome collective action problems by providing important information about the preferences of the parties, monitoring implementation and publicizing failure to comply with commitments, and helping to focus on and work out the principle of fairness. Unfortunately, the existing international institutions rarely provide these benefits, and it is the domestic politics within states that matters most. Moreover, the failure to reform most international institutions during Secretary-General Annan's term of office at the UN has meant that they remain weak, short of resources, poorly staffed (despite some very able officials), and as divided by ideologies and interests as the states themselves.

In short, relying on institutions to do or help to do what states themselves are unable or unwilling to do is not likely to solve the problem of the failure to provide some necessary international public goods. It is not an accident that various proposals to bypass or ignore the current institutional system and create more effective alternatives to it have become increasingly prominent in recent years. One such proposal is the creation of a "league of democracies," but who exactly would qualify is not completely clear, what legitimacy it would have is not apparent, and what actions or policies it would pursue is not self-evident. A different alternative in the environmental arena would create "coalitions of the willing" who would set standards for themselves and not exclude future participation by states willing to subscribe to those standards. We shall return to this discussion in a later chapter. Suffice it to say here that the provision of international public goods is likely to remain intrinsically difficult, unless the internal policymakers in individual states can be persuaded that their interests call for it. One must hope that change can occur short of a disaster that compels attention to the need to cooperate, at whatever cost to the principle of sovereignty or to narrow interpretations of self-interest. It helps to emphasize that the major obstacles are essentially political and moral and that value-laden conflicts can only be resolved or not resolved through the political process.[14]

2.3 Policymaking and Negotiations

There is widespread agreement about the nature of an appropriate policy response by the US (and presumably other developed countries) to the energy and environmental crises. All such approaches seem to agree that the response needs to be comprehensive, that it needs to combine different policies in different time frames, that compromises of various sorts will be necessary, and that we should aim for a system that is resilient and less vulnerable to shocks and surprises. The elements of the combination are obvious: conservation, efficiency improvements, increased spending on alternative fuels, technological innovation, increasing domestic fuel supplies – as far as possible – and diversifying foreign suppliers away from potentially unstable or hostile exporters. There is also agreement that the public understands the dangers of sharp and sudden increases in the price of oil but is unwilling or perhaps unable *as yet* to change behaviors appropriately. Moreover, the wealth and power of the oil and coal companies (and their political supporters) is still vastly greater than the power of alternative fuel suppliers or the environmental lobbies.[15] When one throws into this equation uncertainties on future scientific and technological developments and disagreements about what steps to assign highest priority, the outcome is clear: political immobility, rhetorical posturing, and small incremental movement.[16] One difficulty with generating public support is the very erratic nature of oil price surges and retreats: oil at $4 or more a gallon may galvanize public support, but a relatively quick drop back to $2 per gallon dissipates enthusiasm for change.

The difficulties of getting support for a set of comprehensive, costly, and debatable policies, especially when they require a serious focus on long-run dangers and long-term obligations to future generations, are so severe that some analysts argue the need to refocus the debate on national security.[17] While this may or may not be useful in generating public and political support, it does create some serious difficulties. Security itself is a highly subjective term: how much is enough, who should define when we have met our security needs, and how should we measure its presence or absence? If energy and environmental security are now indivisible, does a focus on national security only generate insecurity or an even more severe outbreak of nationalism or perhaps even imperialism (recall the arguments for using force against oil countries in the 1970s)? Can we seek security only for consumers or must we also include producers? There are also some potential contradictions in the quest for security. For example, the very call for energy "independence" could lead to increased shortages of supplies and higher prices (hurting especially the poor), as could the demand to increase supplies in the strategic petroleum reserves.[18] In short, redefining these issues in terms of security may be helpful, but it is hardly a panacea and could be dangerous if the potential ambiguities are not recognized.

The suggestions about the components of a comprehensive set of policies may be generally useful but they suffer from two major deficiencies. In the first place, they fail to provide much useful guidance about how to achieve stipulated goals; in fact, they are much better at indicating just why those very same policies are unlikely to be politically feasible. Second, even if one supposes that a comprehensive package of measures could be passed, they may fail to provide any guidance about which policy goal should have priority over the others. What should come first and does it make any difference? Presumably, in a policy universe of very scarce resources, one needs some guidance about priorities and/or the principles by which those priorities are set. Simply relabeling the issues as related to security does not suffice.

The negotiating process, which we shall discuss in detail later, has also been caught in the same ideological wars that create domestic difficulties. For example, Michael Grubb, a professor of climate change at Imperial College in London, bemoaned the failure of an earlier conference on global warming: "When something like this is killed, it is killed by an alliance of those who want too much with those who don't want anything."[19] A more recent debate about whether we are approaching an environmental "tipping point" also illustrates the way in which extraneous arguments can deflect attention from more practical matters.[20] Thus negotiations may fail not only because of the usual political and economic hurdles but also because ideology and emotion can influence outcomes. Awareness of the range of obstacles to successful negotiations does not guarantee a better outcome, but may provide insight into the participants' pressures.

We shall discuss possible negotiating strategies in Chapters 7–10, however, a few preliminary comments may be useful at this point. In the first place, the negotiating problem is multidimensional, in effect a multidimensional chess game with different boards, different players at each board, and different tactics and strategies in each game. How to move all the games in the same direction is not clear but there needs to be at least a "sufficient consensus" at each level before a viable final agreement is likely to be feasible. The complexity of satisfying so many interests at the same time suggests a quest for unilateral options, or small group coalitions. It also suggests that "big bang" negotiations at global conferences may not be the best approach and that some form of "gradually accelerating incrementalism" may be more appropriate. Ultimately, success may depend on the convergence of a number of factors: strong political leadership by key countries (not just the US), side payments to losers from any agreement, "variable geometry" so that different countries with different capacities can achieve shared goals at different speeds, and a sense by all actors that all are making credible commitments to implement agreed policies.

Furthermore, one needs to try to make some useful distinctions about the characteristics of different issues and the different degrees of risk associated with different choices. Choucri, for example, notes that there are some environmental processes and outcomes that are outside human control (say, solar radiation), some with partial human control (say, the rapid accumulation of carbon dioxide in the atmosphere), and some entirely or nearly entirely within human control (say, stratospheric ozone depletion).[21] This suggests that we need to focus on issues where state intervention (and the support of international institutions) has the best chance of working to affect individual and group behavior and where many or most states have the capacity, alone or with technical help, to carry out agreed policies. The problem here, of course, is that some of the most dangerous outcomes, such as global warming through carbon dioxide emissions, may also require some of the most extensive, complex, and politically difficult interventions, such as, for example, altering the transportation system and incurring higher costs for electricity. Seeking actions that can be done quickly and that do not require a lengthy negotiation with Congress and other legislative bodies is the kind of small step that makes contextual sense.[22]

Another illustration is Bjorn Lomborg's suggestion[23] that cutting carbon emissions could be accomplished more effectively by sharply increased spending on low carbon technologies, rather than continued efforts to renegotiate a Kyoto Protocol that has failed to achieve its goals. While there is no guarantee that the increased research spending will produce better results, the notion here is rather like the idea of "subsidiarity" in the European Community: do not always seek grand "global bargains" at the highest level but rather focus on the lowest level that can effectively perform a desired task.

Finally, there is a small point that bears repeating. Given the prevailing uncertainties and the unwillingness to take big risks, it is very easy to fall into the trap of doing nothing or of asking for a degree of clarity about topics like global warming or alternative fuels that is unlikely to be forthcoming for some time, or waiting until it is too late or much more costly to act. We must be ready to ask what risks we want to insure against. If those who are warning about the growing risks of global warming and continued dependence on carbon-based fuels are too gloomy, the costs to society will be high but not catastrophic, for we would have expended resources on things that needed to be done anyway, even if at a rate and to a degree that was premature. If on the other hand, those who dismiss global warming or how quickly and extensively it will occur are wrong, the results could be catastrophic for the developed countries as well as for the developing countries, and indeed for the earth itself. Uncertainty should not be used as an excuse for procrastination.

We do not know everything we would like to know, but we know enough to justify serious efforts to slow global warming and reduce our dependence, as quickly as possible, on fossil fuels.

Notes and References

1. Donella H. Meadows, Dennis L. Meadows, Jørgen Randers, and William W. Behrens III, *The Limits to Growth: a Report for the Club of Rome's Project on the Predicament of Mankind* (New York: Universe Publisher, 1972).
2. See Donella H. Meadows, Jørgen Randers, and Dennis L. Meadows, *Beyond the Limits: Confronting Global Collapse, Envisioning a Sustainable Future* (White River Junction, VT: Chelsea Green Publishing Company, 1993), and Donella H. Meadows, Jørgen Randers, and Dennis L. Meadows, *Limits to Growth: the 30-Years Update* (White River Junction, VT: Chelsea Green Publishing Company, 2004).
3. Herman E. Daly, *Beyond Growth: the Economics of Sustainable Development* (Boston: Beacon Press, 1997).
4. Quoted in William D. Sunderlin, *Ideology, Social Theory, and the Environment* (Lanham, MD: Rowman and Littlefield, Inc., 2004), p. 286.
5. Bjorn Lomborg, *The Skeptical Environmentalist: Measuring the Real State of the World* (Cambridge, UK: Cambridge University Press, 1998).
6. See Julian Simon, *The Ultimate Resource* (Princeton, NJ: Princeton University Press, 1983), passim.
7. William D. Sunderlin, *Ideology, Social Theory, and the Environment*, op. cit., is interesting and useful on ideological posturing by both sides in the debate.
8. Jesse H. Ausubel and Paul E. Waggoner, *Dematerialization: Variety, Caution, and Persistence*, (Washington, DC: Proceedings of the National Academy of Sciences, 2008).
9. Jesse H. Ausubel, "The Liberation of the Environment: Technological Development and Global Environmental Change," lecture at the Hungarian Academy of Sciences, 1994.
10. Quoted in John Tierney, "Use Energy, Get Rich and Save the Planet," *New York Times*, April 21, 2009, p. D4.
11. Ibid, p. D1.
12. Michael P. Byron, *Infinity's Rainbow: the Politics of Energy, Climate and Globalization* (New York: Algora Publishing, 2006), p. 3.
13. See Barbara Koremenos, "Loosening the Ties that Bind: a Learning Model of Agreement Flexibility," *International Organization*, Vol. 55, No. 2, Spring, 2001, pp. 289–325.

14. For a similar argument, see J. Samuel Barkin and George E. Shambaugh, eds, *Anarchy and the Environment* (Albany, NY: Suny Press, 1999).
15. See Andrew C. Revkin, "On Climate Issue, Industry Ignored its Scientists," *New York Times*, April 24, 2009, p. A1 and A14.
16. See David Rothkopf, "New Energy Paradigm, New Foreign Policy Paradigm," pp. 187–213, and John Podesta and Peter Ogden, "A Blueprint for Energy Security," pp. 225–39, in Kurt M. Campbell and Jonathan Price, eds, *The Global Politics of Energy* (Washington, DC: Aspen Institute, 2008).
17. See David Rothkopf, "New Energy Paradigm, New Foreign Policy Paradigm," pp. 208–9, ibid., and Thomas L. Friedman, "Show Us the Ball," *New York Times*, April 8, 2009, p. A23.
18. Useful on these questions are Daniel Yergin, "Energy Under Stress," in Kurt M. Campbell and Jonathan Price, eds, *The Global Politics of Energy* (Washington, DC: Aspen Institute, 2008), pp. 27–43, and A. F. Alhaji, "What Is Energy Security?" *Energy Politics*, Vol. 5, No. 2, Spring, 2006, pp. 62–82.
19. Quoted in Andrew C. Revkin, "Odd Culprits in Collapse of Climate Talks," *New York Times*, November 28, 2000, p. F1.
20. Andrew C. Revkin, "Among Climate Scientists, a Dispute Over 'Tipping Points'," *New York Times*, March 29, 2009, p. 3.
21. Nazli Choucri, "Introduction: Theoretical, Empirical, and Policy Perspectives," p. 32, in Nazli Choucri, ed., *Global Accord: Environmental Challenges and International Responses* (Cambridge, MA: MIT Press, 1995).
22. See Daniel F. Becker and James Gerstenzang, "Obama's Power Plays," *New York Times*, April 25, 2009, p. A17.
23. Bjorn Lomborg, "Don't Waste Time Cutting Emissions," *New York Times*, April 25, 2009, p. A17.

3

Surveying the Field

To appreciate the complex issues and opportunities that are becoming so urgent in the 21st century, it should be recognized that "energy and the environment" is one topic, not two. If we wish to limit the threats to our environment, whether local smog or planet-wide global changes, we must face in the end the wide-ranging issues on how we use energy. At the same time, any effort to establish secure future energy sources or to transform our use of those resources in a constructive direction will have to take account of environmental effects for good or for bad.

Setting priorities among a multitude of choices calls for factual material considered with respect to technical feasibility and cost, but any assessment will necessarily involve an admixture with policy, heavily weighted by the

The Challenge of Climate Change: Which Way Now? 1st edition.
By Daniel D. Perlmutter and Robert L. Rothstein.

relative values that the decisionmaker associates with each possible outcome. Because each choice reflects a complex mixture of political, social, and cultural values, there are no neutral or purely technical decisions in this arena.

Energy is a measure of the ability to do work; that is, a measure of the ability to create change in our physical world. The objective of such work could be a huge construction project whose implementation demands extraordinary force and physical strength, or it may be a matter of careful control as in a computer operation achieved without much expenditure of mechanical effort. The goals of our work should be open to social as well as technological criticism, but whatever the objectives in particular cases, all modern life requires us to draw upon energy resources and to adapt them for working applications. We may prefer to reduce or increase our use of energy as we find a new behavior to be more consistent with our preferred lifestyles, but the use of energy to accomplish change is a process that we cannot do without.

In this chapter we begin to address the initial technological steps in the larger process of setting priorities, focusing first on our past and present use of energy with the object of recognizing how change can and did occur. Following that we elucidate the malleable aspects of energy, because an understanding of feasibility in this arena must rest above all on clarity about energy, its many forms, and how they may be transformed for various purposes.

3.1 A History of Change

It may be recalled that one of the major criticisms of the limits to growth argument discussed in Chapter 2 is that it failed to take into account technological ingenuity and market responses to increasing scarcity. Information to partially test such a view may be found in the record of fuels used over time. Such trends may be seen in Figure 3.1, which presents a comparison of fuels used since the middle of the nineteenth century.[1] Examining the estimates shown in the graphical presentation makes it clear that coal gradually replaced wood, that oil then replaced coal, and that after the mid-1970s natural gas began to progressively substitute for some petroleum uses.

Are these changes to be attributed to market forces or to the "ultimate resource," human technological ingenuity? Since wood, coal, and oil were abundantly available during the century after 1850, it is safe to give credit to market forces for much of the change as users moved from one fuel to the next in response to price changes and consumer desire for convenience. To

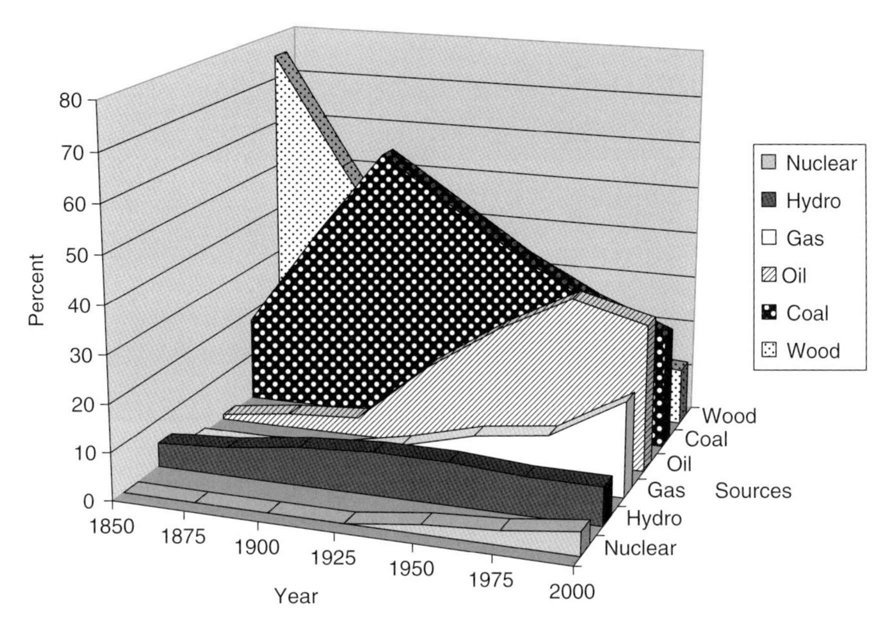

Figure 3.1 Changes in world energy sources.

be sure there was significant invention involved: among other things, furnaces had to be redesigned to accommodate each new fuel, strip mining for coal was developed, and new drilling techniques invented to improve recovery from existing wells or to create access to new oil deposits; however, the technological changes that were called for were largely extensions of prior engineering ideas, hardly radical innovations.

In contrast to earlier history, any extrapolations to the future depend on the likely growth of nuclear power as well as new technologies that are truly radical and do demand human ingenuity. It may be noted that the presentation in Figure 3.1 is entirely silent with respect to newer developments such as solar energy or power from wind, waves, or tides. We will revisit these alternatives when the discussion turns to renewable resources and possible new energy sources.

Even more pertinent to our current concerns is the fact that in the past the transitions from one fuel to the next were not politically or economically controversial, because neither scarcity of resources nor environmental issues were of great concern at that time. When changes gave some sector of the economy an advantage over another, its necessity was taken for granted in the larger

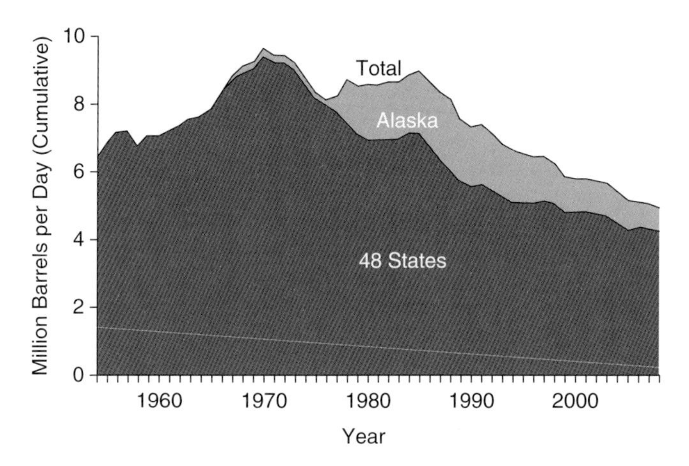

Figure 3.2 US crude oil production.
Source: US Department of Energy, Energy Information Administration, Annual Energy Review 2008

picture. Ecological damage, such as resulted for example from uncontrolled strip mining, was largely ignored over most of the period and only noticed at all as a local effect. Lacking sharp controversies, the changes could be made without creating the kinds of stresses implicit in adopting an entirely new approach to energy supply and use. Our current situation will in several respects require much more radical change in a context of uncertainty, and should be expected to generate disagreements backed by strong opinions.

More immediately, it is instructive to examine the well-documented experience of the major producers of oil. US crude production[2] peaked in 1970, as shown in Figure 3.2, and the non-OPEC (Organization of Petroleum-Exporting Countries) countries as a whole were at a peak or plateau in production between 2004 and 2008.[3] The introduction of Alaskan sources improved the picture, but the goal of independence from imported oil has thwarted each succeeding administration and the US has been increasingly dependent on foreign imports. In 1970 under the leadership of then President Nixon, who spoke out in favor of energy independence, the country still imported 22% of its petroleum needs; by 1991 during the presidency of the first President Bush, the imports had grown to 42%; and by the year 2007, the figure was at 72%. At the 2009 price of oil, this latter level of imports costs some $700 billion per year and has a very significant impact on the nation's balance of trade.

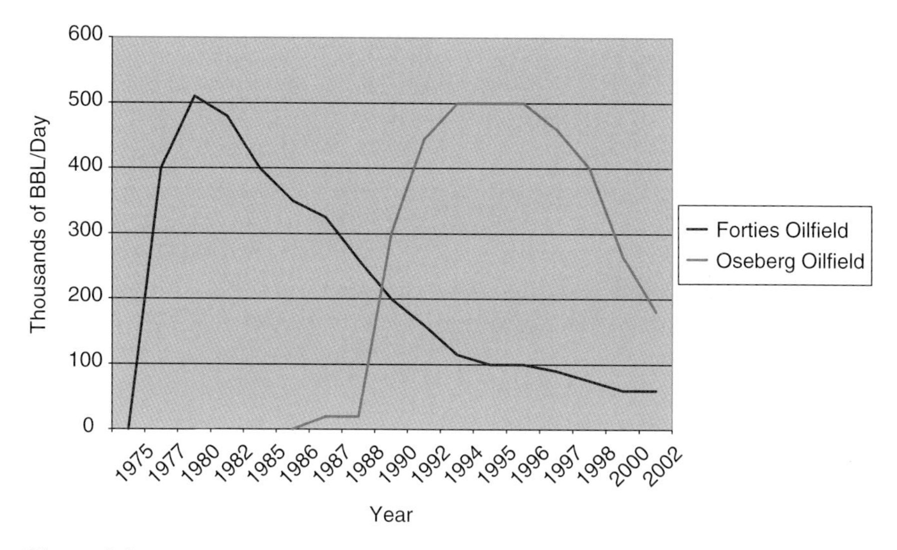

Figure 3.3 Crude production profiles from two giant oilfields.

But what of the pattern for individual oil fields? Data on crude oil production over the years from two giant and three super-giant oilfields are given in Figures 3.3 and 3.4, in units of barrels per day. There are 42 gallons in a barrel (abbreviated BBL). While these records from the largest oil pools in the world differ in their details, most striking is the trend to a maximum rate of production that drops off as the various wells begin to be depleted. This means, in effect, that a continuing supply at an undiminished rate depends on: (i) finding ways to access any oil that might remain in the depleted wells; (ii) repeatedly discovering new finds to compensate for the established wells as they run dry; or (iii) exploiting hitherto less attractive alternative sources.

Each of these possibilities has been considered in a study published by the International Energy Agency (IEA),[4] with the results summarized in Table 3.1. It is clear from the IEA analysis that from about 2010 on we will have a diminishing supply of oil from currently producing fields and will have to depend more and more heavily on the non-conventional alternatives (such as tar sands, for example), or on fields not yet developed or not yet found. By their nature, dependence on undeveloped or unfound sources must introduce discomfort from the residual uncertainty, and the IEA projection indicates that a huge part of what will be needed fits into these uncomfortable categories.

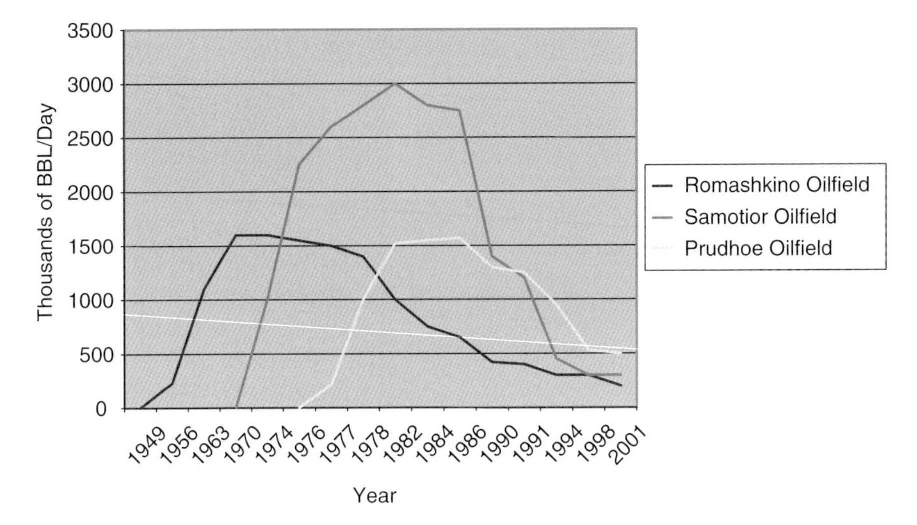

Figure 3.4 Crude production profiles from three super-giant oilfields.

Table 3.1 World oil production scenario for 2030.

Source of Oil	*Production Rate (Million BBL/day)*	*Percent*
Crude Oil-Currently Producing Fields	27	26
Crude Oil-Fields Yet to Be Developed	23	23
Crude Oil-Fields Yet to Be Found	20	19
Natural Gas Liquids	20	19
Non-conventional Oil	8	8
Additional Enhanced Oil Recovery	5	5
Sums:	103	100

3.2 Measuring Energy

In order to compare and contrast the many energy options that have already been proposed in both the public media and technical publications, as well as those suggestions that are yet to come forth, we need to have a scale of measurement of energy. Without such a scale it would be impossible to distinguish large from small effects, the trivial from significant. As with consideration of any quantitative subject, there are multiple scales in common use. Even a relatively familiar, simple measure such as distance can be expressed in feet, yards, and miles, or in centimeters, meters, and kilometers; not to mention the more recent focus on molecular dimensions in nanometers.

Because the electrical unit of a watt is widely in use for domestic lighting and for monthly utility bills, it is a familiar and useful measure of the rate of energy use. The 100-watt light bulb is well known in all developed societies and in fact became an icon representing ideas and progress for much of the last century. If we consider a typical small home to simultaneously need light from ten 100-watt bulbs, the rate of energy usage is ten times 100 watts; i.e., 1000 watts, also called a *kilowatt*. In a small neighborhood of a thousand such homes, the rate of usage would be 1,000 kilowatts, called a *megawatt*. The rate of energy use, that is the energy use per unit time, is what is meant by the term *power*. When it serves the purpose, the quantity of energy can be recovered from that of power by multiplying the rate of use by the time period of its use. This measure of energy is often expressed in *kilowatt-hours* (kwh).

To go one step further with the urban geography illustration, a medium-size city of a million homes would require 1,000 megawatts of power, a *gigawatt*. Of course, the full demand for electricity is much greater than that required just for light bulbs, including myriad appliances, and the demand can fluctuate wildly at different times of the day or year. A well-known example is the notorious sudden call for increased power in the early evening hours when workers return home to use air conditioners, dryers, and dishwashers. While kilowatts, megawatts, and gigawatts are useful to approximate demands for homes, towns, and small cities, larger cities and whole countries need to be scaled with yet larger units. When this is needed the unit is the *terawatt* (abbreviated TW), equal to 1,000 gigawatts. A mnemonic illustration for these steps is shown in Figure 3.5.

3.3 Supply: Where Do We Get It?

Before going on to consideration of new options, it is instructive to review the *status quo*, since it is the existing supplies of energy that we seek to supplant or at least reduce in usage. How are we meeting current needs? Data

One Bulb	100 watts
Ten Bulbs One Home	1,000 watts = Kilowatt = kW
Thousand Homes One Town	Megawatt = MW
Million Homes One City	Gigawatt= GW
Billion Homes One Country	Terawatt = TW

Figure 3.5 Measurement of electric power.

Table 3.2 Sources of US Energy, 2008.

Source	*Use in TW*	*Percent*
Petroleum	1.24	37
Natural Gas	0.80	24
Coal	0.74	23
Nuclear	0.26	9
Biomass (including Ethanol)	0.14	4
Hydroelectric	0.07	2
Geothermal + Solar + Wind	0.03	1
Sum:	3.28	100
Sum in GW = 3,280		

from 2008 are available from the US Department of Energy (DOE) [5] and are summarized in Table 3.2. The entries in the table are averages over extended time intervals, since the actual rates of energy use fluctuate with time and place. The numbers demonstrate that 93% of use in the US came from just four sources: oil, coal, gas, and nuclear. Of these, petroleum is the overwhelming leader at 37% and nuclear is the smallest at only 9%. The use of

Table 3.3 Sources of World Energy, 2006.

Source	*Use in TW*	*Percent*
Petroleum	5.8	33
Coal	4.3	24
Natural Gas	3.6	21
Biomass (Wood + Crop Waste + Dung)	2.0	11
Hydroelectric	1.1	6
Nuclear	1.0	5
Geothermal + Solar + Wind	0.05	0
Sum:	17.8	100
Sum in GW = 17,800		

natural gas to generate electricity has increased in the last few years, and it now exceeds coal as a major resource. Of the remaining sources, biomass (wood, corn for fermentation into ethanol, crop waste) and hydroelectric power account for about 4% and 2%, respectively, and geothermal, wind, and photovoltaic solar energy are collectively less than 1%, virtually negligible in the big picture in 2008. Note that the power amounts listed in the table are expressed in terawatts to avoid a scale that would call for many zeroes, but the last line shows the sum in units of gigawatts.

A similar audit is available with respect to world-wide usage,[6] complete to the year 2006. Petroleum is still the leader at 33%, but coal, wood, and hydroelectric power are more significant in the wider world than they are in the US (Table 3.3). Hillring and Parikka[7] have emphasized that to meet their self-imposed targets, the nations of the European Union could move to greater utilization of wood from plantations, forest and agricultural residues, and by-products of the wood-processing industries. It is interesting to find that together, crop waste and animal dung make a contribution to the world supply that is greater in size than that of nuclear power; these agricultural sources may diminish as living standards improve in developing economies. The listings in Table 3.3 also make it easy to compare the world total at 17.8 TW with the total US figure of 3.3 TW, putting out front the fact that the US alone accounts for almost one-fifth of the world-wide rate of energy consumption. This observation has led to criticism of US policy from some conservationists, but the intensity of such complaints may be softened as the consumption of energy among the developing nations is increased by their shift to rapid growth.

3.4 Demand: How Do We Use It?

In the US, energy consumption is about equally divided among three uses: transportation, industry, and domestic heating. Up to now transportation has been sharply dependent not only on the source of energy but especially on its form; that is, our cars, planes, and trains are run on liquids of high chemical energy content per unit volume. We need such concentrated energy to achieve the desired accelerations and range of travel between fills of our gas tanks. At present, gasoline, diesel fuel, and jet fuel come from petroleum, with some allowance for ethanol in gasoline, but fuels of the future may come from quite different sources. Among the possibilities that are to be discussed in subsequent chapters are coal or biomass gasification to make synthesis gas (a mixture of hydrogen and carbon monoxide). This first step can be followed by selected catalytic reactions to form hydrocarbon or alcohol liquids. Such developments would provide the liquid fuels that we need while freeing us from dependence on petroleum.

Vehicles that use electricity for power are also being developed as petroleum-free alternatives for transportation. By using high-capacity batteries to store energy, vehicles would ultimately get their power from whatever fuel is used in the generation of electricity. Currently some 95% of US generating plants use coal, gas, nuclear, or hydroelectric sources, thus avoiding petroleum. The detailed breakdown is given in Figure 3.6.[8]

The source of energy in industry varies greatly, depending on whether the production processes are primarily mechanical, electrical, or chemical in nature. Numerous examples can be found of factories that use each form of energy. Manufacture of aluminum metal, for example, calls for huge investments in electric power, as does chlorine manufacture. Coal converted to coke is a major ingredient in smelting ores to refine metals. Mechanical force is essential in forming or rolling metals. The entire industry that produces or uses plastics and artificial fibers is a major user of chemicals as either reactants or catalysts. This almost endless list offers many opportunities for a reduction of amounts or replacement of one energy input by a more advantageous form.

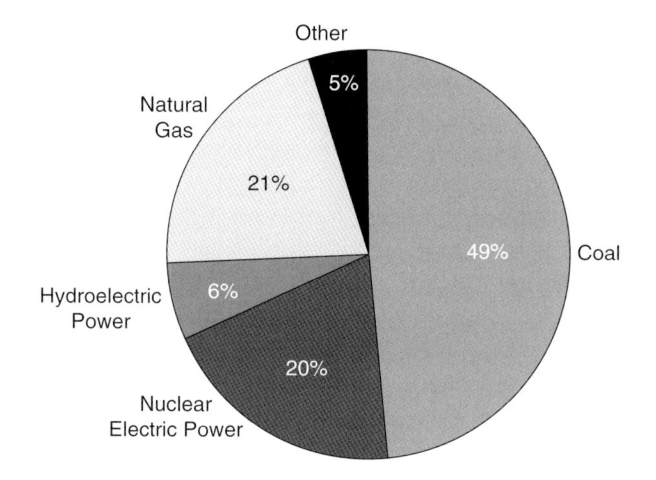

Figure 3.6 Fuel sources for US electricity generation, 2008.
Source: US Department of Energy, Energy Information Administration, Annual Energy Review 2008

Home hot water and space heating is still primarily based on oil and gas, although new developments include solar hot water heating. Home electricity use, whether for heat pumps, air conditioners, or appliances depends on whatever energy source the local power company uses in its generating plants. Today, depending on location, the sources can be any fossil fuel, hydroelectric, or nuclear. In fact the energy is likely to be coming from all these origins simultaneously, if the power is being drawn from a shared wide area grid. In the future, domestic needs may possibly be met by fuel cells individually installed in each home, but these are not yet commercially fully developed. If and when available, such an alternative would provide greater local flexibility and might possibly in the long term move use to fuels that do not depend on petroleum sources.

3.5 Will We Run Out of Oil? Or Gas?

Since petroleum is today our major energy source and is expected to remain so into the near future, there is for planning purposes an obvious and essential question to be asked: how long will it last? The answer is a serious domestic matter for the US and has major international consequences.

There are estimates available from a number of sources[9] of world-wide proved oil deposits, as well as estimates of deposits not yet found but expected to be found. The latter are, of course, very difficult to pin down, because they depend on so many subjective factors. Recall that quantities of petroleum are reported as barrels of 42 gallons each (abbreviated BBL). The lowest estimates of the world-wide ultimate recoverable resource (URR) are at about 2,000 billion BBL, in part extrapolated from the US experience of a peak in 1970. More optimistic numbers come from the US Geological Survey (USGS) which in 2000 offered an estimate of 3,345 billion BBL. We also know quite closely our current rate of use, annually about 30 billion BBL. Are these facts not sufficient to make the needed estimate of our future? They would be except that the current usage is bound to increase as world populations and their physical demands grow. Looking forward some two decades to 2030, we might project an expected world demand of say 40 billion BBL per year. These figures give us brackets on the time remaining, depending on the specific assumptions made: for the low URR estimate and our projected usage, we can anticipate having oil for at least the next 50 years; if the high URR is closer to reality, we can expect about 82 years of supply. These simple calculations are summarized in Table 3.4, where the spread in the results arises from the levels of uncertainty in the estimates, as well as the degrees of optimism in the assessments. A compromise between the extreme views may be found in a recent study from the UK Energy Research Centre, which surveyed more than 500 publications to obtain and evaluate 14 forecasts of world oil production. Their report[10] focused on when peak production

Table 3.4 Will we run out of oil?

	Proved World Reserves	*Ultimate Recoverable Reserve: URR = Low*	*Ultimate Recoverable Reserve: URR = High*
Estimate (Billions of BBL)	1,000	2,000	3,300
Current Usage (Billions of BBL/year)	30	30	30
Projected Usage in 2030 (Billions of BBL/year)	40	40	40
Years Left at Current Usage Rate	33	67	110
Years Left at Projected Usage Rate	25	50	82

of conventional petroleum should be anticipated and concluded that 2030 was the most likely date. Of course peak production is not the end of the line, and we should expect that the supply line will not at that date run dry overnight.

Additional uncertainties with regard to this matter arise from a number of directions, some technological, some economic, and some political. Of these, even if the details of its development can be complicated, the technology is the most straightforward in its immediate objectives: the geology of oil deposits need to be made more reliable, new drilling techniques will need to go to deeper deposits, and improved chemical engineering will be needed to extract oil from shale, tar sands, and other reluctant donors. Whatever is technologically achievable at any given time is an important determinant in how much petroleum is available as proved reserves; that is, how much can be extracted from the ground.

The major economic uncertainty in assessing proved reserves is the price of petroleum on the world market. As price increases, suppliers may be more willing to produce and sell what they have, drillers more willing to drill deeper and travel further to inhospitable sites. At least as important is the sensitivity to competition among the potentially higher priced "unconventional" alternatives. For example, some 1.2 million gallons per day of oil was profitably extracted from tar sands in Alberta, Canada when the price of petroleum briefly spiked to $145 per barrel during 2008. That made Canada the biggest foreign oil supplier to the US, accounting for 19% of imports that year,[11] and yet the factories threatened to close down when the competing petroleum crude fell in price to $45 per barrel.

Clearly, the oil that is available and extractable at any given time is a function of its selling price, but it is also the case that price changes in response to demand. The conservation strategies of the 1970s Carter administration, for example, were successful in reducing consumption until the price of oil dropped sharply during the 1980s. The great fluctuations in price over the three decades from 1980 to 2008 are shown in Figure 3.7, expressed in then current dollars.[12]

The political uncertainties around oil have to do largely with its uneven distribution. As shown in Table 3.5, much of the world supply is located in places of conflict and/or political instability, either current or likely in the near future. Not all sources should be viewed in this light, however. Actual countries of origin for US imports of petroleum in 2008[13] are given in Figure 3.8, showing that Canada and Mexico are among the largest suppliers.

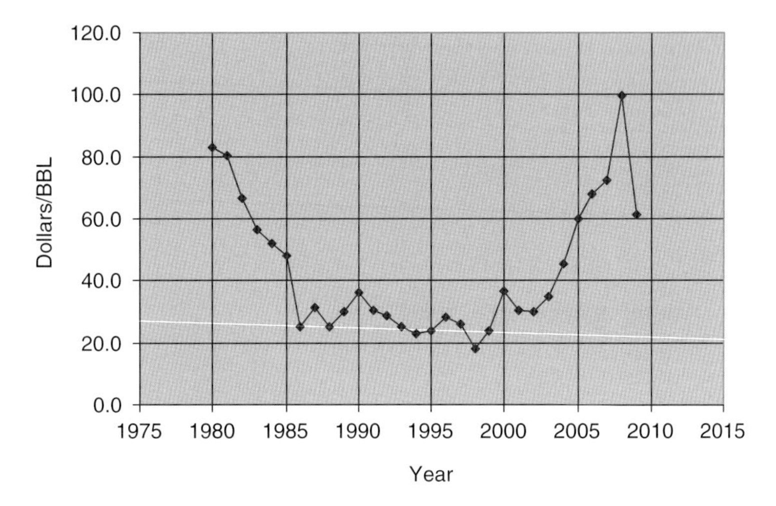

Figure 3.7 Petroleum prices from 1980 to 2008.

Table 3.5 Locations of petroleum deposits.

Region	*Percent*
Middle East	65
South America	9
Africa	7
Former USSR	6
North America	6
Asia Pacific	4
Europe	2
Total	100

Information on amounts in various locations are released by interested parties, whether governments or private holders, who may believe that the effect of their statements may be more important than their accuracy. Furthermore, a proved reserve is only a true reserve if it is able to be produced and shipped to the consumers, and we have seen on more than one occasion that supplies are cut off for various reasons of conquest, war, piracy, rebellion, or national interests.

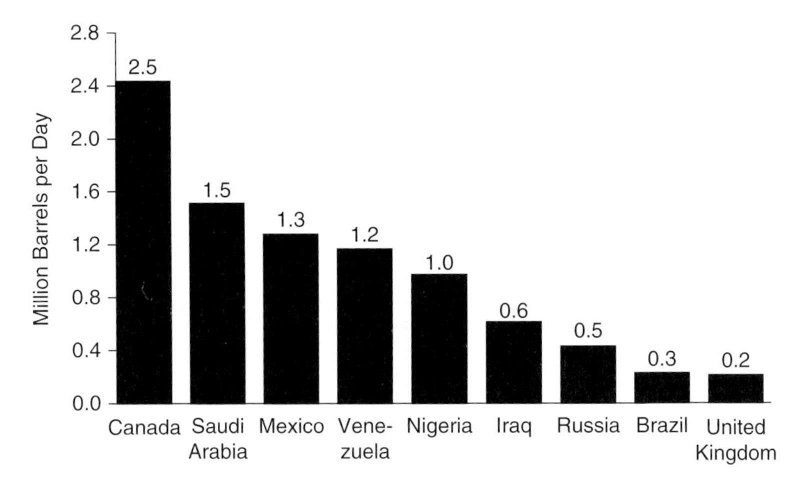

Figure 3.8 US petroleum imports, 2008.
Source: US Department of Energy, Energy Information Administration, Annual Energy Review 2008

Because a number of oil-producing nations have based their national budgets on the expectation of selling petroleum at high prices, they face internal crises when prices fall sharply. A recent estimate[14] listed per barrel oil prices at which six nations would just manage to break even: Qatar at $24, Libya at $47, Saudi Arabia at $55, Nigeria at $65, and Iran and Venezuela at $90. While it is true that the oil producers need to sell petroleum as much as the consumers need to buy it, the perceived short-term political gains, domestic or between nations, can on occasion override longer term economic interests. Conversely, long-term commitments can have damaging economic consequences if there are rapid short-term changes. In 2009, for example, Russia's Gazprom was forced by a prior long-term contract to import natural gas from Usbekistan, even though Russia then had to sell the gas at a loss.[15]

Finally, even if a usable estimate is made of what is and will be available in the foreseeable future, the forecast on annual use is still not entirely unambiguous. As the developing world becomes more prosperous, its appetite for petroleum increases. In recent years the growth in China and India has been especially noteworthy, but the issue may be expected to have even greater impact as nations in Southeast Asia, Africa, and South America make economic progress. Given such a forecast, it appears that the projected annual world usage of 40 billion barrels that led to the larger estimate of 82 years' supply may in fact be much to optimistic, and the reserve could be used up in an appreciably shorter period. To balance the picture properly, however, it should be added that new discoveries of petroleum still continue to be made in many parts of the world.

Any treatment of fossil fuel supply and demand must also consider the mixture of hydrocarbon gases that have come to be called *natural gas*. As it comes out of the ground this gas is composed primarily of methane, with smaller amounts of other flammables (ethane, propane, butane, pentane) and some impurities. After purification steps this gas has become the second most important source of energy in the US, and it is exceeded only by coal in the world-wide assessment of energy consumption. As with petroleum, any estimate of the extent of gas deposits is subject to uncertainties of geology, technology, economics, and politics; nevertheless, figures are available from which it is possible to talk about reserves and rates of use. The Energy Information Administration of the DOE[16] has compiled an estimate of US recoverable natural gas, as of January, 2007. The overall figure is 1,747 trillion cubic feet, but it is striking that of this total only 211 trillion cubic feet are classed as *proved reserves*. The rest consists of *unproved* probabilities in four different categories: (i) undiscovered, (ii) inferred, (iii) gas dissolved in petroleum, and (iv) unconventional sources. Clearly each of these four is associated with appreciable uncertainties. A comparable world total for January, 2009 is available from the same source: 6,254 trillion cubic feet.[17]

As we did above for petroleum, we may ask how long these resources might last if they are to be used at the rate of current consumption. The results of these simple ratio calculations are summarized in Table 3.6, indicating that the world proved reserves can offer an assured supply for some 60 years, but that the current US proved reserves might be used up in only nine years. Of course the number based on proved reserves is the most extremely pessimistic possibility, since it does not allow for ongoing exploration and new discoveries. In fact recent findings in the US have been unexpectedly large, especially those associated with shale deposits, and the country's reserves have jumped by 35% in just two years. This dramatic increase in US domestic supply is from so-called *unconventional sources*, a category that includes tar sands and other combined materials that require more expensive extraction than that

Table 3.6 Reserves of natural gas.

	US Proved	*US Total Estimate*	*World Proved*
Reserves (Trillions of cubic feet)	211	1,747	6,254
Annual Consumption (Trillions of cubic feet/year)	23.2	23.2	104
Years Left	9	75	60

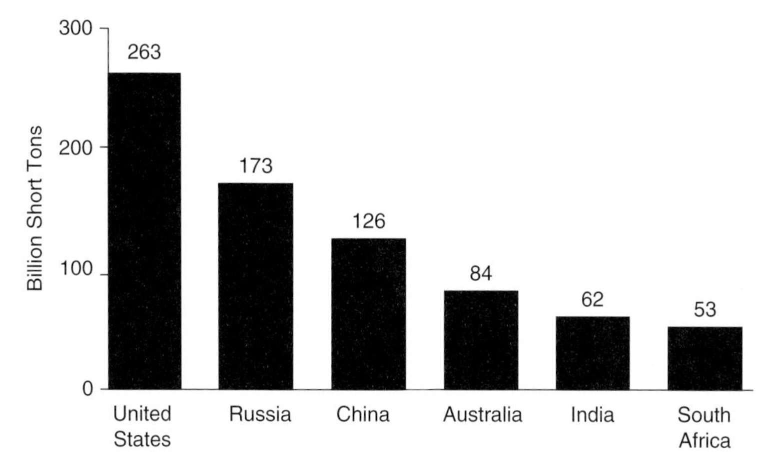

Figure 3.9 World recoverable reserves of coal, 2005.
Source: US Department of Energy, Energy Information Administration, International Energy Outlook 2009

following conventional drilling. If instead of proved reserves we were to use in our calculation the more likely figure of the US total recoverables, the assured duration increases dramatically to 75 years.

With respect to coal, the third of the major suppliers of fossil energy, the world recoverable reserves are estimated to be about 929 billion tons, widely distributed but mostly located in the US, Russia, and China (see Figure 3.9 for 2005 data).[18] If current consumption at 6.5 tons per year were not to change, this supply would last 142 years. The US portion of the total reserves is 260 billion tons, currently being used at a rate of 1.1 billion tons per year.

In summary it appears safe to say that we have adequate fossil fuel supplies to meet current needs for at least the next half century, but that the picture may change if the world-wide demands expand even more than anticipated. But this having been said, it remains to ask whether we *want* to continue burning fossil fuels in light of our current understanding of what the effects are on our world climate and all that depends on it. It seems that we have at least a window of opportunity, a period of time during which if we choose to do so we can continue to burn fossil fuels in order to make possible a transition to a different technological and economic lifestyle.

3.6 Forms of Energy

As a preliminary to any examination of energy alternatives, it is essential to recall one underlying truth: that energy is neither created nor destroyed, only changed in form. It is in a way similar to buying a new currency in exchange for what one already has. After the exchange, appearances are different but the underlying value remains the same. Any study of options is therefore a study of transformation from one form to another, and the best we can ever hope for is that the new energy form will be more available, more flexible, or in general more useful in meeting our needs. This notion needs to be further elaborated by recognizing what in fact the various forms of energy are.

Each of us encounters energy in various forms in everyday circumstances. *Radiant energy* from the sun reaches us out of doors on any clear day. We see part of this energy directly as visible light and sense the invisible part of the sun's spectrum as heat when it is absorbed in our skin. The sun's radiation is also the source of the photosynthetic reactions that cause plants to grow; thus, the radiant energy is transformed into *chemical energy* where it is stored in the plants until they are eaten, burned, or decay. If humans or other animals eat the plants, the energy is stored in yet other chemical forms in the animal tissues.

Over eons of geological time some of the then existent plants and animals were buried and chemically altered when they were subjected to extreme conditions of temperature and pressure. Those chemical changes formed the coal, oil, and gas we use today – the so-called fossil fuels. In burning them we are in essence releasing the sun's energy which was so long stored in those fuels in the form of chemical hydrocarbons. Our purpose in burning the fuels is, of course, to release this stored energy in the form of *thermal energy*, which can be used to generate steam under pressure. Until the steam pressure is released we may think of it as having potential for work not yet done. When it is used to drive a steam engine or a turbine wheel, that potential is realized; that is, the *potential energy* is converted into *mechanical energy* of motion, also called *kinetic energy*. In turn, the turbine drives a dynamo to generate electricity.

A similar transformation of potential into kinetic energy takes place when stored water behind a dam is released through a turbine. The spinning motion of the turbine is also used to generate electricity, giving us what we usually call hydroelectric power. Tracing back the steps in this transformation, we find that it all started from the sun, when radiant energy evaporated water that condensed into rain and snow. The water ran off into rivers that filled the liquid volume behind the dam. Subsequent release of the potential in the stored water made it possible to generate *electrical energy*.

It is worth noting that there are many variations on this theme. In the device called a fuel cell, for example, the chemical energy in a fuel is directly converted into electricity, without the intermediate steps. In the internal combustion engine that drives our cars the chemical energy in gasoline (or ethanol) is converted into mechanical energy to move the automobile. In the newer hybrid cars some of the chemical energy of the fuel is used to charge a battery; that is, to store energy in another chemical form. In addition, when the brakes are applied, some of the kinetic energy in the wheels drives a small generator that also charges the battery.

Commercial nuclear reactors can be viewed in this context too, since they are in essence devices for boiling water to form steam for use in generating electricity. They depart from the more mundane generating plants in that they use *nuclear energy* as their source for transformation into thermal energy. A condensed summary of the various forms is given in Table 3.7.

Finally, it should be noted that electricity is not necessarily the end of the line of transformations. If the electric power is dissipated in the heating elements of a toaster, it is for the purpose of liberating thermal energy; if it feeds power to a microwave oven, the electrical energy is converted into electromagnetic waves that in turn heat food to a desired temperature.

All the foregoing may be generalized by observing that the various forms of energy can be converted from one form to another. Some of these transformations were part of nature long before humans came into the picture. Other steps are part of the technological revolution of the last few hundred years. A brief selection of such conversions is offered as Table 3.8.

The fact that energy can be transformed in form should not be taken to mean that all transformations are equally efficient. For some steps the

Table 3.7 Forms of energy.

Kind	*Examples*
Radiation	Sunlight
Chemical	Plants, oil, gas, batteries
Thermal	Heat
Potential	Steam pressure, water behind dam
Kinetic	Motion of water wheel, windmill, automobile
Mechanical	Motion of machinery
Electrical	Current in wires
Nuclear	Uranium fission, hydrogen fusion

Table 3.8 Energy conversion examples.

	Energy Conversion	
Example	*From*	*To*
Plant Photosynthesis	Radiation	Chemical
Burning Coal	Chemical	Thermal
Steam Pressure	Thermal	Potential
Turbine, Steam Engine	Potential	Mechanical
Dynamo (Generator)	Mechanical	Electrical

efficiency of transfer is very high indeed: the dynamo, for example, converts the kinetic energy of motion to electricity at a rate better than 98%. On the other hand, conversion of heat into mechanical motion is typically only possible to the extent of about 30%. In that case the 70% that cannot be converted must be rejected as heat at a lower temperature. An internal combustion engine typically converts less than 25% of the chemical energy in the fuel into motion of the wheels; the balance is again rejected as heat from friction or into the radiator coolant (water or glycol) and from there to the ambient air.

It is the recognition that energy can be converted from one form into another that led ultimately to the historical observation that energy is neither created nor destroyed, only transformed. To that should be added, however, that in any transformation from one form to another, some fraction of the energy is degraded in quality and may emerge in a form less able to serve our needs; in that sense it can be said of energy conversion that there is no free lunch. The comparison to currency exchange is again useful here, if one includes in the value assessment the fees that are charged by the banks or agents that perform the exchange. These themes of transformation and loss will arise repeatedly in what follows, and with these limitations in mind we will in the chapters to come explore a diverse set of ideas for improving our energy status.

Notes and References

1. US Department of Energy, Energy Information Administration (EIA), *Annual Energy Review 2008*, published June, 2009, Figure 5, p. xx, http://www.eia.doe.gov/aer/pdf/aer.pdf.
2. Ibid., Figure 5.2, p. 130.
3. Matthew R. Simmons, *Twilight in the Desert: the Coming Saudi Oil Shock and the World Economy* (Hoboken, NJ: Wiley, 2005); also Energy Information Administration (EIA), "World Primary Energy Production by Source, 1970–2006," International Energy Database table, www.eia.doe.gov/aer/txt/ptb1101.html.
4. International Energy Agency, *World Energy Outlook 2008*; also reported in Richard A. Kerr, "Energy: World Oil Crunch Looming?" *Science*, Vol. 322, November 21, 2008.
5. US Department of Energy, EIA, *Annual Energy Review 2008*, op. cit.
6. Energy Information Administration (EIA), "World Primary Energy Production by Source, Section 11" International Energy Database table, www.eia.doe.gov/aer/txt/ptb1101.html.
7. Bengt Hillring and Matti Parikka, "Potential Market for Bio-Based Products" in P. Ranalli, *Improvement of Crop Plants for Industrial End Uses* (Netherlands: Springer, 2007), pp. 509–21.
8. US Department of Energy, EIA, *Annual Energy Review 2008*, op. cit., Figure 8.2a.
9. Richard A. Kerr, "Splitting the Difference Between Oil Pessimists and Optimists," *Science*, Vol. 326, November 20, 2009, p. 1048, www.ukerc.ac.uk/support/Global%20Oil%20Depletion.
10. Ibid., p. 1048.
11. Jad Mouawad, "Report Weighs Fallout of Canada's Oil Sands," *New York Times*, May 18, 2009.

12. US Department of Energy, EIA, *Annual Energy Review 2008*, op. cit.
13. US Department of Energy, EIA, *Annual Energy Review 2008*, op. cit., Figure 5.4.
14. Jad Mouawad, "OPEC Plans Further Output Cut," *New York Times*, December 16, 2008.
15. Andrew E. Kramer, "Falling Gas Prices Deny Russia a Lever of Power," *New York Times*, May 16, 2009.
16. US Department of Energy, Energy Information Administration (EIA), *Annual Energy Outlook 2009*, Report No. DOE/EIA-0484.
17. US Department of Energy, Energy Information Administration (EIA), *World Proved Reserves of Oil and Natural Gas*, www.eia.doe.gov/emeu/international/reserves.html.
18. US Department of Energy, EIA, *International Energy Outlook 2009*, op. cit.; also US Department of Energy, EIA, *Annual Energy Review 2008*, Figure 11.13.

4

Global Warming

The common understanding of how our use of energy affects the environment has in the last few decades undergone major changes. Even before the formation of the Environmental Protection Agency (EPA) during the Nixon administration, the odors and visible particles in the air over large cities in the US were recognized to have detrimental effects on public health. The advent of lead-free gasoline was in recognition of the hazards resulting from burning gasoline that contained tetraethyl lead. In short, our collective consciousness, as it were, accepted the idea that smog and air pollution are unpleasant and sometimes harmful, and that measures were needed to reduce the hazards.

The Challenge of Climate Change: Which Way Now? 1st edition.
By Daniel D. Perlmutter and Robert L. Rothstein.

Table 4.1 EPA criteria air pollutants.

Pollutant	*Primary Source*
CO (Carbon Monoxide)	Incomplete combustion
NO_2 (Nitric Oxide)	Formed in engines and furnaces
SO_2 (Sulfur Dioxide)	Product of coal combustion
PM-10 (Particles)	Soot formed in diesels and fires
Lead	From leaded gasoline, flaking paint
O_3 (Ozone)	Formed in air from NO_2 and hydrocarbons

In response to this recognition, the EPA put the various pollutants into the specific categories that are shown in Table 4.1 and set standards for each of the air pollutants; maximum concentrations were established for each of the pollutants in order to keep health risks to acceptable levels. The specific limits that are set by the EPA are not always met, and it should also be noted that over time the EPA limiting maximum concentrations tend to be reduced as better information becomes available on the degree of each hazard. The ground level limit on ozone, for example, was set at 0.084 ppm (parts per million) in 1997 during the Clinton administration. During the subsequent Bush presidency the standard became 0.075 ppm, and a new EPA proposal[1] set forth in 2010 seeks to limit ozone to 0.06–0.07 ppm to be phased in over two decades. Positive improvements have been recognized, but vary considerably since each of the pollutants has its own sources and each is removed from the air by a distinct mechanism.

The most successful action in response to the EPA expectations has been to reduce lead in the air by eliminating the use of tetraethyl lead as an additive to gasoline. Further improvement in air quality was achieved by inserting catalytic converters into the exhaust pipes of automobiles, as well as adding clean-up treatments to power plant effluents. Together, these changes reduced nitric oxide and sulfur dioxide emissions. These two pollutants form acid droplets in the atmosphere and can be identified as acid rain when they fall to the surface as part of ordinary precipitation. Ozone, one of the most directly irritating components of the air, is not released as such but rather forms in the air via secondary chemical reactions with nitric oxide and hydrocarbons. Accordingly, its effects can be mitigated by reducing the concentrations of its precursors. It should be noted that although fallout such as acid rain sometimes affects larger catchment areas, the most severe effects of most of the pollutants are usually local in and around particular cities, never global in reach.

Such air pollution issues are still with us, but in addition we are now faced with a potentially more serious problem: that of global warming. This new concern is far reaching in two respects. First, it will to a greater or lesser extent affect everyone, everywhere on earth. Secondly, its effects are predicted to be catastrophic in at least some places, and once in place they are virtually irreversible over a human lifetime. Moreover, as Aldy and Stavins and many others have noted, dealing with global warming and climate change is the ultimate global commons problem, because gases remain in the atmosphere for decades and they mix uniformly in the upper atmosphere "so that damages are independent of the location of emissions."[2] And, if these prospects are not frightening enough, the remedies being considered call for major changes in our current lifestyles, changes not easily accomplished when the damage will only be recognized as occurring in a distant time and place.

In recognition of these larger scale considerations the EPA has widened its portfolio as to what is identified to be unacceptable emissions, and in 2009 the agency proposed to impose nationwide limits on emissions from power plants, refineries, and other large facilities to the extent that they contribute to global warming. This chapter will elucidate the details of these concerns and arguments, and begin to consider the framework in which some resolutions may be sought.

4.1 Temperature of the Planet

From time immemorial, long before any animal life evolved on our planet, the sun has been heating planet earth, and yet except for some ups and downs along the way, the temperature has leveled off. Evidently, there is a compensatory process that allows the collected heat to be lost, otherwise the temperature would have risen to levels that would not have supported human life. How does the energy transfer take place?

Recalling that only half of our planet is in daylight while the other side is in night makes it clear that all the incoming radiation heats only one hemisphere at a time; the other side gets no incoming radiation and can act as a source of radiation out into space. Overall these two are in balance: the incoming energy and the outgoing must be equal for the temperature of the planet to stabilize. The figures are known: overall, the energy reaching our planet from the sun is 1,353 watts over each square meter (W/m^2), but about one-third of this radiation is reflected back by clouds and other surfaces, producing an average temperature of the planet at −18°C = 1°F. That is to say, if our world were viewed from space very far away, it would appear to be uncomfortably cold, well below the freezing point of water. This result is so different from our everyday experience that it calls for an explanation.

We on earth are very fortunate to have a protective atmosphere that provides where we live an average surface temperature of about 15°C (58°F), well above the planetary average and in the range where water is liquid. Of course there is still wide variation of temperature depending on location, time of day, time of year, and weather, but the average at the surface is well above that of the atmosphere as a whole.

To understand how our atmosphere is protective, it is essential first to recall that in addition to its visible light solar radiation includes invisible components (called ultraviolet) that are too energetic to be detected by our eyes. The air is transparent to both the visible and the ultraviolet rays; that is, they penetrate the atmosphere, come to the earth, and heat the surface. When, during the cooling phase of the cycle, the surface radiates outward, it does so as heat waves (called infrared); however, the longer wavelength heat is absorbed by a number of components in the atmosphere, notably water vapor, carbon dioxide, ozone, nitrous oxide, and methane. As illustrated in Figure 4.1 this process produces a blanket over our planet, in many respects similar to the heating in a greenhouse, and these components have come to be called *greenhouse gases* (GHGs).

4.2 Greenhouse Gases

It is so nice to live in a warm place. If the greenhouse gases keep us warm and protect us from what would otherwise be destructive cold, why do we fear them? It is a matter of having too much of a good thing, and we fear the bad consequences of overheating. The relative influence of each of the greenhouse gases depends on both its concentration in the atmosphere and its ability to absorb infrared radiation.

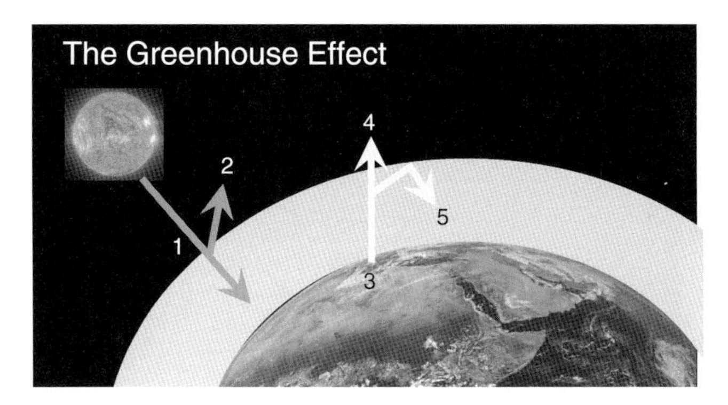

Figure 4.1 The greenhouse effect (source: National Oceanic and Atmospheric Administration (NOAA) Paleoclimatology Section, http://www.ncdc.noaa.gov/paleo/globalwarming/what.html). 1, Some solar radiation passes through the atmosphere and heats the Earth's surface. 2, Some of the solar radiation is reflected back into space. 3, Earth's heated surface emits infra-red radiation (heat waves) toward space. 4, Some of the infra-red radiation passes through the atmosphere and some is absorbed by greenhouse gases. 5, Some of the energy absorbed by the GHGs is re-emitted back to Earth and heats the lower atmosphere and the surface.
Source: National Oceanic and Atmospheric Administration, Paleoclimatology Section

Water as a vapor in the air is a major contributor to the greenhouse effect. But water is also present in the atmosphere in the form of liquid droplets that form clouds. In contrast to greenhouse heating, cloud cover can have a contrary effect by providing a reflective surface for incoming sunlight. The overall effect of water is thus a combination of heating and cooling that varies in time and place as precipitation and humidity change. Since these changes are weather dependent, their influence is largely beyond our control. Suggestions have been put forth recently that it might be possible to change the cloud cover in a constructive direction. The possibilities of such geoengineering will be discussed in a later chapter.

Ozone is an important GHG competitor to carbon dioxide. It is present in the atmosphere at both low and very high altitudes. It is formed at high altitudes (in the stratosphere) from ordinary oxygen by exposure to ultraviolet radiation from the sun, and it serves as a radiation filter that protects life on the surface of our planet. It was widely recognized in the 1970s and 1980s that this protective role was being compromised, because the ozone was being removed by chemical reaction with chlorofluorocarbons (CFCs) released by several human interventions. A conference was convened in Montreal in 1987 and a protocol was agreed upon to remedy the problem by reducing CFC production and substituting less damaging compounds for use in refrigeration, air conditioning, foam processing, and pressurized containers. The response from industry was rapid and effective: a new family of chemicals was synthesized, tested, and sold to replace the offending CFCs. The new compounds are designated as HCFCs (hydro-chlorofluorocarbons) to indicate that hydrogen atoms have been put in the place of some of the halogen atoms in the CFCs. The success of the Montreal Protocol and its amendments in 1992 are hallmarks of how international agreement can work to everyone's benefit.

Ozone is also formed by chemical reactions involving nitric oxide and hydrocarbons at low altitudes where we live (in the layer of the atmosphere called the troposphere). It is one of the pollutants that the EPA has long sought to control, because it is especially obnoxious in large cities where traffic density provides the reactive hydrocarbons in the exhaust from automobiles and buses. Although ozone is a potent GHG, it is chemically reactive and is relatively short lived in the troposphere. If it is not regularly replaced, it spontaneously disappears from the atmosphere in a matter of days to weeks and cannot be a long-range threat.

Besides water vapor and ozone there are a number of other gases in the atmosphere that qualify as GHGs, at concentration levels that have been changed and are still changing because of human intervention. The most famous of these is carbon dioxide (chemical formula: CO_2), which we add to every day by burning fossil fuels in large quantities. The growing awareness of the carbon dioxide burden has put the focus on this gas as the prime offender, and various nomenclatures have been coined to describe its influence. We speak of *carbon emissions*, a *carbon footprint*, a *carbon-based economy*, and *carbon permits*; meaning in each case to refer to carbon dioxide gas.

While lesser attention has been paid to some of the other greenhouse gases, they also can have significant effects on climate. Those of greatest concern are listed in Table 4.2, together with their current concentrations and their estimated lifetimes in the atmosphere. This last item of information is critically important, since it tells us whether our actions will affect the lives of our children and grandchildren. All the gases in the list are very long lasting in the atmosphere with the possible exception of methane, the only entry that has a life expectancy measured in decades.

The CFC refrigerants are also potent GHGs, but as reported above, steps have already been taken in recent years to eliminate their effects by substituting less damaging working fluids in our refrigerators and air conditioners,

Table 4.2 Greenhouse gas concentrations in 2005 and their lifetimes.

Gas	*Atmospheric Concentration (ppm)*	*Lifetime to Reduce Concentration to One-Third*
Carbon Dioxide (CO_2)	379	Century
Methane (CH_4)	1.8	Decade
Nitrous Oxide (N_2O)	0.3	Century
CFCs	Various Very Low Concentrations	Years to Centuries
HCFCs	Various Very Low Concentrations	Years to Centuries

and by collecting the leftover chemicals for recycling or destruction. Some compounds in this family are persistent over many years, others are especially short lived. The figures range from a low of 45 years (for CFC-11) to a high of 1,700 years (for CFC-115). Fortunately, the concentrations of these compounds in the air are extremely low, typically only about one-thousandth of the carbon dioxide concentration. The HCFC group is of lesser concern because of its still lower concentration in the air as well as its less damaging greenhouse potential.

The presence of methane in the air arises from a variety of sources. The greatest fraction comes from ruminant barnyard animals (cows, sheep, water buffalo) via enteric fermentation that takes place in their digestive systems. The second largest methane input is attributable to sloppy housekeeping practices in the production and transmission of fossil fuels, especially natural gas and petroleum. This latter issue has become the focus of conservation efforts recently in response to an information release from the EPA estimating that 3 trillion cubic feet of methane are leaked into the air each year by the oil and gas industry.[3] As shown in Figure 4.2, the prime offenders are Russia and the US, followed by Ukraine, Mexico, and Iran. These numbers are contested by the Russian state gas monopoly, admitting to a number only half as large; on the other hand, the EPA review implies that all these figures may be too low. Other releases of methane result from rice cultivation, coal production, wastewater disposal, landfills, and residential combustion of biofuels.

There are yet other possible sources of methane that are difficult at this time to assess, but are potentially very large. These deposits of methane are loosely bound up in layers of permafrost and as hydrates deep under the cold

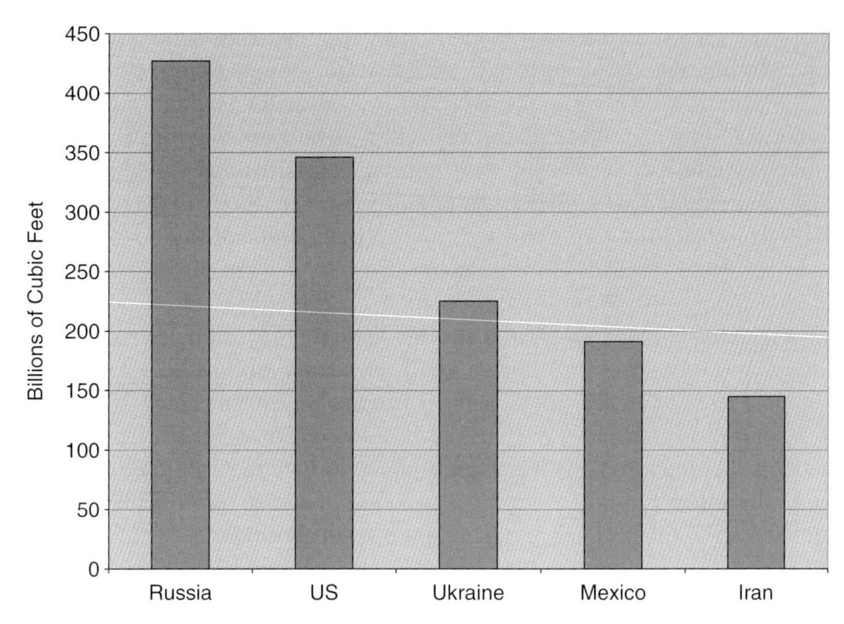

Figure 4.2 Methane emissions from oil and gas industries, 2006.

oceans. They are held in place as long as their surroundings remain cold enough, but the methane can be freed to escape to the air if temperatures rise appreciably. There is a threat of a snowballing effect in this picture, because as newly released methane enters the atmosphere due to global warming, it would act to create further warming, in effect a positive feedback loop that feeds upon itself. The extent of this hazard and its sensitivity to global temperature is at this time not well defined.

Next in line as a GHG is nitrous oxide (chemical formula: N_2O, also commonly called laughing gas). Globally, about 26 million tons of this gas were released to the air each year during the 1990s, of which approximately one-third was anthropogenic in origin. This human contribution is primarily from fertilization of soils, combustion, and industrial processes. Because of its very low concentration in the air, the effect of this release on global warming is at this time relatively small, but the emissions of N_2O are increasing at a rate estimated to be in the range of 0.2–0.3% each year.[4] Furthermore, this gas is only eliminated from the air quite slowly, requiring about a century

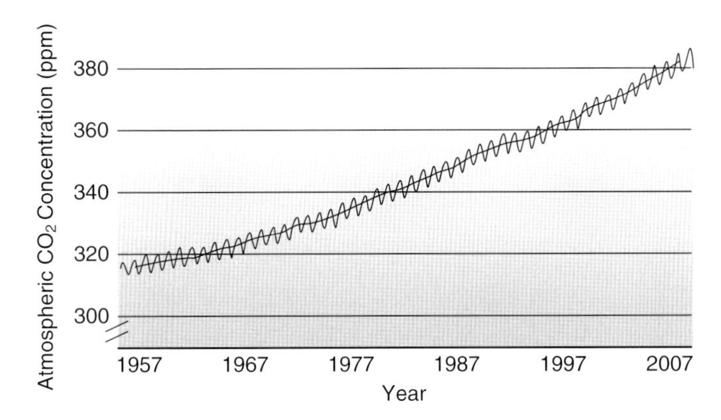

Figure 4.3 Carbon dioxide measurements, 1958–2007.
Source: Monthly mean atmospheric CO_2 at Mauna Loa Observatory, Hawaii, graphed by Hugo Ahlenius, UNEP/GRID-Arendal

to lose half its concentration. Because of this, the accumulated effect may be troublesome if the rate of release continues to grow through the twenty-first century.

Given all these considerations, CO_2 still remains the focus of attention, and because CO_2 is the prime player in global warming, it is vital to examine the record to see if its concentration in the atmosphere has changed in recent years. Reliable data are available going back to 1958. In graphed form it is sometimes referred to as the Keeling curve[5] in honor of the chemist who developed the measuring techniques. As is evident in Figure 4.3, the concentration has increased by about 20% over the last half century, reaching a level of 379 ppm in the year 2005, and continuing to rise at a rate of about 2 ppm each year. The periodic fluctuations in the concentrations are caused by the seasonal changes during each year, as plants absorbs CO_2 during their growth period and as winter furnaces burn more fuel to support their longer periods of heating use.

Furthermore, there is clear evidence that the planet is warming, in spite of the climate fluctuations that occur in many small time intervals. The record since the middle of the 19th century,[6] shown in Figure 4.4, emphasizes the sharp increase in global average temperature that occurred following the onset of the industrial revolution, the period during which the burning of fossil fuels steadily increased. The surface measurement data came from a wide range of land- and ocean-based stations, presented here with the zero level as the mean

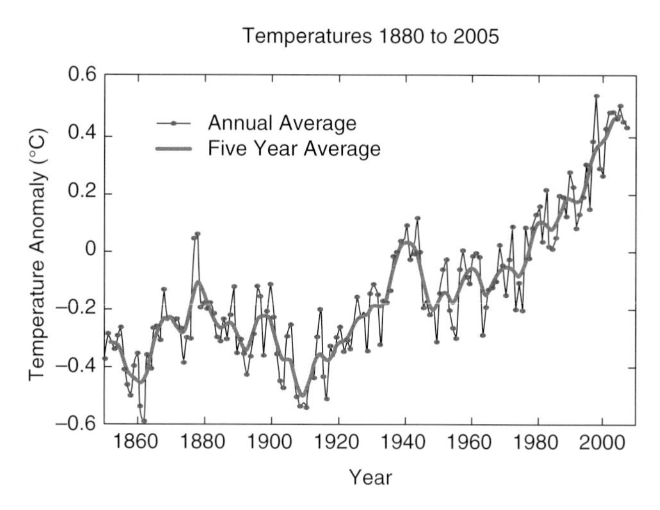

Figure 4.4 Global temperatures, 1880–2005.
Source: IPCC Report, Climate Change 2007: The Physical Science Basis

temperature between 1961 and 1990. While the increase appears at first scan to be relatively small, about 0.8°C (1.4°F), this change is significant in global terms and a harbinger of larger changes to come. It should be noted that there is much noise in the record over any small interval of time, because ocean heating and cooling cycles produce annual fluctuations that are large compared to the long-range changes. Clear trends show up when the data are averaged over 5–10-year periods. An assessment of surface temperature measurements released by NASA (National Aeronautics and Space Administration),[7] for example, reported that the decade from 2000 to 2009 was the warmest on record. Their data covering the past 30 years showed a trend in temperature rise of 0.2°C (0.4°F) per decade.

Increases in temperature over some geographic sectors are thought to be associated with drought, melting ice, hurricanes, population migrations, and other ecological and human catastrophes. The evidence for melting ice is strong, and connecting events such as hurricanes with ocean heating is strongly suggestive, but not yet fully agreed upon by all experts in these fields. The most detailed state-of-the-art hurricane model for the Atlantic Ocean region predicts that there will be more of the most intense hurricanes, although the total number of storms will be less.[8] Huge human migrations have been caused by drought and floods in various parts of the world, but have so far

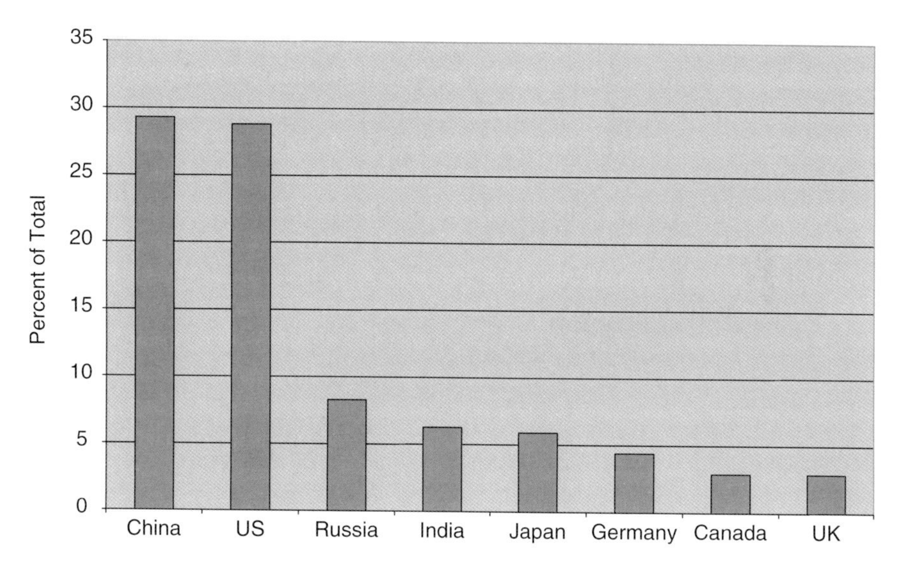

Figure 4.5 Percentage carbon dioxide emissions by nations, 2006.
Source: US Department of Energy, Energy Information Administration, Annual Energy Review 2008

not been directly attributed to global warming. Yet, if the associations are correct, we will pay a high price indeed in the future for ignoring them today.

Since they are so intimately tied to fossil fuel combustion, the sites of CO_2 production are not uniformly distributed around the world. US Department of Energy (DOE) data[9] for 2006 show Chinese CO_2 emissions at 29% of the world total, just barely exceeding that of the US, but the exceptional growth of manufacturing in China has called forth a demand for energy that surpasses all other nations, and it appears that by 2008 China was clearly in first position world-wide. In 2006 the US was a close second with about 28% of the CO_2 emissions, with Russia, India, and Japan following, as is shown in Figure 4.5. Because two nations account for such a large part of the total emissions, it is widely felt that no international agreements aimed at reducing emissions of CO_2 are possible without agreement between the two giants, the US and China. In the words of one of the US climate negotiators[10]: "Certainly no deal will be possible if we don't find a way forward with China." What then are the difficulties?

The impediments to agreement on the control of emissions stem above all on fears that any restrictions on energy use will interfere with economic development and growth. In the bilateral negotiations between the two giants the arguments that are presented by each side are based on a variety of views, each claimed as equitable by its proponent: (i) as a latecomer to industrial growth, China would like the freedom that the US enjoyed for the past century and a half; (ii) China is emitting four times as much CO_2 as the US for every unit of gross domestic product; (iii) the Chinese emissions are less per capita than those of the US, which has a smaller population; (iv) China's emissions are growing faster than any other major country; and (v) most of the CO_2 that has already entered the atmosphere came from past behavior of today's wealthy nations, not from China. However, in spite of these formidable obstacles in negotiation, there is agreement from all sides that something must be done to limit long-range effects, and that assurances are needed that future emissions will be "measurable, verifiable, and reportable." One study by the Electric Power Research Institute[11] estimated that, lacking new controls, the current trend would lead to a CO_2 level in the atmosphere of 450 ppm by the year 2070.

4.3 Is Global Warming Our Fault?

What facts about global warming can we cite with certainty? First that carbon dioxide is definitely a greenhouse gas, and its concentration in the atmosphere has dramatically increased over a period of many decades. Given that the timing of the increase corresponds so closely with the industrial growth that led to huge increases in fossil fuel combustion, it is reasonable to attribute most of the increased concentration to this change. To put all this in a longer time perspective, however, it should be said that geologists have demonstrated by sampling relics of bygone eras that the planet had undergone temperature ups and downs long before humans had any input, and it is fair to ask how much of the more recent change is caused by anthropogenic (human behavior) contributions.

An answer to this question is to be found in the reports of the International Panel on Climate Change (IPCC), a scientific intergovernmental panel established in 1988 by the United Nations and World Meteorological Organization. In their Fourth Assessment Report[12] entitled *Climate Change 2007*, the panel concluded that: "Most of the observed increase in global average temperatures since the mid-20th century is *very likely* due to the observed increase in anthropogenic greenhouse gas concentration." The phrase "very likely" is defined in the report to mean that expert judgment assessed the likelihood to be over 90%. This panel report represents a consensus among hundreds of authors and editors, and was reviewed by more than 620 experts and governments. The panel findings are based on a variety of scientific measurements that include the melting of glaciers in many parts of the world, heating of the oceans, and the retreat of sea ice, as well as atmospheric and surface temperatures. A review by the US National Academy of Science agrees with their results, and the American Physical Society 2007 statement on global warming said that warming was "incontrovertible" and could lead to significant ecological and social disruptions. An excellent summary of the history and current status of the subject was the subject of the 2009 AAAS (American Association for the Advancement of Science) Presidential Address by James J. McCarthy.[13]

Support for this IPCC finding was also voiced recently[14] by John Holdren, President Obama's science advisor, who has been quoted to say that "any reasonably comprehensive and up-to-date look at the evidence makes clear that civilization has already generated dangerous anthropogenic interference in the climate system." He told a congressional hearing on this subject[15] that "global climate is changing in highly unusual ways compared to long experienced and expected natural variations." By 2009, two years after the most recent IPCC report, the worst-case IPCC projections were being realized. The co-chair of the Copenhagen Climate Congress[16] in March of 2009 told delegates that "emissions are soaring, projections of sea level rise are higher than expected, and climate impacts around the world are appearing with increasing frequency."

The consensus among the vast majority of the scientific community is that the report of the IPCC panel is substantially correct. To the extent that divergent views exist at all, they have to do with the time periods over which particular temperature rises will occur and with the severity of the global changes that will follow. The worst-case scenarios anticipate floods, hurricanes, and droughts, followed by social upheaval in consequence of massive population shifts.

4.4 The RF Index

For some purposes it is useful to have a measure of how effective any particular greenhouse gas is in changing the energy balance of the planet.

Specifically, we may ask: what is the current difference between the energy carried by incoming radiation and that of outgoing radiation, and how has the difference changed in response to some stimulus? If this change is the result of an increased GHG concentration in the atmosphere, we can use it as an index of the relative effect and importance of that GHG as a potential climate change mechanism. For this comparison the computed change is based on an early date prior to the industrial era. The name given to this index is *radiative forcing* (abbreviated RF), measured in units of watts per square meter (W/m^2). The change in equilibrium surface temperature is proportional to the RF: a positive value of RF indicates a warming of the climate, a negative value tells us that we are moving toward cooler surroundings. As an example of its application, an RF of 3.75 from a doubling of CO_2 concentration would lead to a surface temperature increase of 3°C.

The IPCC report cited above has supplied a detailed listed of the RF indices for the long-lived GHGs, and the data are reproduced in Table 4.3. Comparing the numbers it is clear why CO_2 is the center of attention and why methane and nitrous oxide are getting increasing study. The effects of the gases controlled by the Montreal Protocol are still present. The major CFC is decreasing by about 1% each year, but its 100-year lifetime means that it will be a factor in global warming for years to come. At this time it is the third most important GHG, as shown in Table 4.3. The HCFCs that were introduced to achieve the objectives of the Montreal agreement also make a noticeable RF contribution, but the effect is limited. Further, the HCFCs of greatest indus-

Table 4.3 RF indices of long-lived greenhouse gases, 2005.

Gas	*RF of Gas Component (W/m^2)*	*RF of Gas Component Relative to CO_2*
Carbon Dioxide (CO_2)	1.66	1
Methane (CH_4)	0.48	0.29
Nitrous Oxide (N_2O)	0.16	0.10
CFCs Total	0.27	0.16
HCFCs Total	0.04	0.02
Others	0.02	0.01
Sum:	2.63	

trial importance have lifetimes of 20 years or less and their concentrations should decrease unless large releases occur in the future.

The last entry in Table 4.3, "Others," refers to a mix of trace ingredients, mainly SF_6 (sulfur hexafluoride), CF_4 (carbon tetrafluoride), and C_2F_6 (hexafluoro-ethane). While at low concentrations, the RF index for these compounds is only about 1% of the value for CO_2; however, it is important to note that they are potent greenhouse gases and have lifetimes of millennia. In effect they make a permanent contribution to the overall RF index. They are products and by-products of power distribution equipment and aluminum manufacture, and their concentration in the air has been increasing linearly for about 40 years.

Because methane's life in the atmosphere is a matter of decades, it is possible to categorize it as intermediate rather than long lived. This view has been adopted by Jackson,[17] who has used the RF index very effectively to argue that in addition to concerns over long-range changes, the effects on climate of medium- and short-lived pollutants need to be addressed separately. Examining for example the RF estimates for emissions (primarily methane) coming from seven major production sources, she showed that significant contributions to global warming can be expected from these inputs over the next 20 years, even though methane's duration in the atmosphere is measured in decades rather than centuries. Jackson pointed out further that these trends would not have been recognized if only long-lived CO_2 and N_2O had been followed, and concluded that two separate treaties are called for to address these different needs. The magnitudes of some of her estimates and their distribution are shown in Table 4.4. The total RF of 0.48 W/m^2 agrees with other figures for methane.[18]

4.5 Air Pollution Revisited

To begin this chapter we distinguished air pollution from global warming, since the former was local and essentially a matter of public health while the latter was global and a more general hazard than personal and communal health alone. But there is clear evidence that the same aerosol particles, sulfur dioxide, and ozone that play a major role in influencing individual health also affect earth's radiation balance. Now we can see that in fact the two areas have more than a superficial connection, for the emissions of pollutants in

Table 4.4 RF indices for short- and medium-lived emissions.

Sources	*RF* *(W/m²)*	*Percent*
Enteric Fermentation	0.14	29
Gas Production	0.09	19
Rice Cultivation	0.07	15
Coal Production	0.06	13
Human Wastewater Disposal	0.05	10
Landfills	0.04	8
Residential Biofuel Combustion	0.03	6
Sums	0.48	100

the form of methane, nitrous oxide, nitric oxide, black carbon (soot), and CFCs all take part in the process of climate change. Some of these are short or medium lived in the air, but even they act as GHGs for a time interval large enough to effect some undesirable global warming. Furthermore, their RF effects are not simply proportional to pollutant emissions, and will of course persist as long as they continue to be released. This analysis is supported by very detailed studies carried out by Shindell and his co-workers,[19] who used established RF values and computational climate models for a variety of local pollutants to argue that policies that target local or regional air quality should also take into account long-term effects on climate. Parrish and Zhu[20] have also addressed this matter, asking how climate change mitigation can be dovetailed with efforts to lessen the health impacts of air pollution.

All this notwithstanding, CO_2, because of its very long life in the air is a special long-term global hazard and will continue to be the focus of all efforts at amelioration. A US Supreme Court decision in 2007 instructed the EPA to decide whether climate-altering gases were a threat to human health, and if so to act to regulate them. This has been formally recognized by the EPA,[21] and the agency now considers control of emissions of GHGs to be within its purview and area of responsibility. By issuing its finding on this matter the EPA has signaled that it could use its rule-making power to control GHG emissions from large stationery sources such as coal-burning power plants and cement kilns, even without further congressional action.

4.6 Immediate or Short-Term Remedies

Before considering ways to limit carbon dioxide emissions, it is essential to place them in the context of an overall dynamic balance; that is, to recognize that the disposition of the gas is part of a series of natural processes that continuously and globally remove the gas from the atmosphere. Trees and all plants that use photosynthesis for growth and maintenance feed on CO_2 and remove it from the atmosphere; the oceans and many soil types are also major sinks as they dissolve the gas to form soluble carbonates and bicarbonates. Given this perspective, it is disturbing to find that major world forests are being destroyed in response to the economic pressures for expanding agriculture. Such behavior not only reduces the global ability of our planet to absorb CO_2 from the air, but also immediately releases huge quantities of the gas as the wood is burned. Rain forest destruction is estimated to account for about 20% of all GHG emissions globally.[22] It is obvious that any remedy to this problem will require major changes in policies within particular countries and among countries. Proposals are under consideration: the REDD (Reduced Emissions from Deforestation and Forest Degradation) program, for example, would pay countries for preserving natural assets.

On the other side of the GHG ledger, a great variety of possible actions have been proposed to reduce our profligate release of carbon dioxide into the atmosphere. Some are technological fixes, others use the pressure of economic advantage to influence behavior, and still others rely on domestic or international political constraints. There are controversies emerging for each of these, because many of these proposals require huge capital investments or are based on treaties or technologies that are as yet unproven. We will examine these ideas in greater detail in subsequent chapters, but take up here the remedies that are attractive and available immediately or in the short term.

While it is true that any movement will entail costs or reduced income to some factions, who will be opposed, attention to greater efficiency and reduction of waste will often lead to improvements that require relatively low capital investment and provide pay-back in relatively short times. We will now focus on opportunities for improved efficiency of use by considering conservation in four sub-categories: transportation, industry, residential, and generation of electricity.

Until recently, modern automobile engines used less than 20% of the chemical energy available in the gasoline used. The remaining 80% was wasted when it was lost to the ambient air. More efficient engines were not welcomed because they were associated with smaller sized cars and slower acceleration of the vehicle, but satisfactory gains were available with diesel engines, more popular in Europe than in the US. Now, with the advent of hybrid technology, engines use less gasoline by being run under more ideal steady conditions to charge the battery that operates the car. All-electric plug-in cars are on many drawing boards but are some years away of being practical. When and if they become reality, the burden of improving efficiency will have shifted from the vehicle to the power plant that generated the electricity in the first place.

For many years the US has been behind other industrialized nations in requiring automobiles to use fuel more efficiently. At the same time that the US numbers have been at 27 miles/gallon for cars and as low as 22 miles/gallon for SUVs (light trucks), the European Union set their standard at 43 miles/gallon and Japan was up to 46 miles/gallon. Even the late-comer China was ahead of the US at 36 miles/gallon. The result of this reluctance to mandate engines that used less gasoline was to be expected: whereas US oil consumption increased 20% in the 27-year span after 1980, virtually all the nations of Western Europe reduced their consumption during that period, some by over 30%. The comparative details are given in Figure 4.6.

If the US had followed the European lead in this direction, it would have taken away the permissive treatment that was extended to SUVs by putting them in the light truck category. This change could have saved an estimated 5% of gasoline use if it got only half of the SUVs off the road. Further, a significant shift to diesel engines in half of the fleet could have saved another 10%, but unfortunately neither step was taken. By 2009 the US Congress was finally ready to take steps that should have been introduced several decades earlier. Federal rules that will begin to take effect in 2012 and be fully in force by 2016 mandate increases in fuel efficiency standards for US cars and light trucks to 35.5 miles/gallon, significantly higher than the requirements of the last 25 years (which were slightly over 25 miles/gallon).[23] At the same time that the US is moving in this direction, other nations are also increasing their expectations for automotive fuel efficiency. Based on proposals being reviewed by the Chinese auto industry, one estimate[24] is that they will require a fleet average of 42.2 miles/gallon in 2015. It should go without saying that any program that creates massive use of public transportation will accomplish even more than efficiency improvements for individual drivers. Where avenues are wide and clear lanes are provided, rapid transit bus systems have

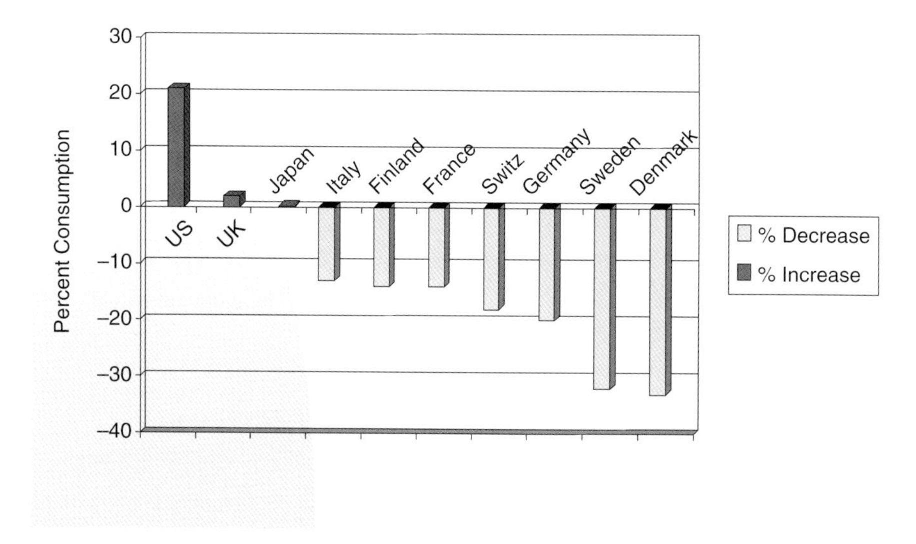

Figure 4.6 Trend in oil consumption, 1980–2007.

been able to move high-density populations out of their commuting cars. One estimate is that CO_2 emissions in Bogota, Colombia, for example, have been cut to about 25% of what would otherwise be the auto expectation.[25]

Industrial applications and electric power generation have in common the need for high temperature, high pressure steam, in the former case primarily as a heat source and in the latter application to drive turbines linked to generators. The steam cools as it gives up its potential energy for either of these purposes, and although the somewhat cooler steam still has useful energy, it was thought in past years uneconomical to bother to recover some of this heat, even though it has been estimated to be about 20% of the potential. Now the picture has changed, and the newest installations find ways to recover heat from the degraded steam, either by further power production, or by using it for secondary purposes such as preheating process feed streams, space heating, or to generate domestic hot water.

In a very different direction there are changes that aim not so much for efficient use of a given fuel, but rather on changing fuels with the object of reducing the CO_2 released. This substitution occurs because for a given release of its chemical energy each fuel has a characteristic but different production of CO_2. Taking oil (from petroleum) as a comparative standard, the CO_2 release from coal combustion is about 30% greater and that from wood is 50% greater for a given energy production. In contrast, the production of CO_2 from the combustion of natural gas is about 24% less, which has prompted movement from coal to oil to gas whenever feasible. This remedy is, of course, dependent on adequate supplies of oil and natural gas at competitive prices, and may only be a temporary measure over the long run.

Since the use of energy in residential and commercial buildings is primarily for heat, light, and to run appliances, efficiency is immediately to be found in better insulation, lighting that consumes less wattage for each unit of light, and more effective space heating. Some of these changes can be implemented easily and some are already in place: fluorescent lights or LEDs (light-emitting diodes) can replace incandescent bulbs, and refrigerators can be bought that are better insulated from their surroundings. Other remedies are architectural in form and require structural changes that are quite feasible in new structures but not so immediately easy to install as retrofits.

Renewable solar energy has been widely used with success, by capturing rooftop sunlight to supply hot water and in less common applications to generate electricity. These topics will get further treatment in the chapter on renewable energy that follows.

4.7 Limits to Growth and the Commons Revisited

In our prior consideration of the limits to growth arguments, we attempted to put into perspective the opposing positions. On the one hand were the warnings against unfettered growth of population that would lead to exhaustion of resources. Opposition came from both those who wanted to see industrialization for third world nations as well as from optimistic economists who expected market forces to innovate and substitute for resources when they were in short supply. But these arguments, whatever their source, were presented at a time before global warming was a serious concern. In our newer context of the twenty-first century many of the former opinions are still voiced and may even be valid, but they are overshadowed by the glaring impact of global warming. It is not merely that we may run out of a particular raw material or fossil fuel, but rather that the very use of a fuel or a chemical process may hasten our descent into unacceptable world-wide chaos.

As the evidence for global warming becomes stronger, it is increasingly difficult to deny its current and future impact. In spite of that there are claims made on the commons by various interest groups and nations, who make the argument that their needs for development are paramount and that others should shoulder the costs of any remedial action. Nevertheless, there are now voices raised in support of international collective action in the name of protecting our planet. It should be noted that climate changes are not distributed uniformly over all parts of the world. The melting of glaciers, for example, though they obviously only occur in selected places can have both local and world-wide effects. The city of El Alto in Bolivia that has long used a nearby glacier as a store of water is now losing this water supply. The effect on the population of some million people is overwhelming and most will have to relocate unless some emergency action is taken. Even more frightening in this regard, a World Bank report found that the loss of Andes glaciers would threaten nearly 100 million people over the next 20 years.[26] Another location that is especially sensitive to global warming is in low-lying Bangladesh, where nearly a fifth of the country is less than 1 m above sea level.[27] Recent forecasts estimate sea level rises of up to 2 m by the end of this century;[28] if this occurs many millions of the population will lose their agricultural lands and their livelihoods. The forced movement of so many people will, of course, have international repercussions.

Increasingly we appear to be placing our hopes on innovation and substitution, using legislation to provide subsidies and economic incentives and depending on market forces to develop alternatives to fossil fuels. Whether these modifications will sufficiently wean our societies away from CO_2 production remains to be seen, but they surely appear to provide an acid test for the economic model based on market responses.

In Chapter 2 we introduced the notion of public goods and the idea of a collective commons. An atmosphere that contains a lower concentration of CO_2 may be thought of as a public good, since its advantages are available to all and no one can be excluded from enjoying its benefits. It is perhaps analogous to the ocean fisheries that cross national boundaries, but is different with respect to what is added or removed from the commons. Free riders in the fishing industry are those who *take* from the oceans while ignoring the need to replenish the supply of fish, whereas the free riders with respect to global warming are those who *add* to the burden of greenhouse gases, refusing to contribute to international limitations.

Putting increasing amounts of CO_2 into the atmosphere is a fine example of the use of the commons, for each violator can feel that the air is public property that he is entitled to use and that, in any case, his individual contribution is small. If enough participants behave in the same way, the cumulative effect can be catastrophic, indeed the "tragedy of the commons." Leaving aside such deviations from good international citizenship, we are left with the need to negotiate the relative responses of all stakeholders, even if there are no free riders and no misuse of the commons. Given that populations all over the world are represented by national governments, how is the burden of compliance to be distributed? Furthermore, how much time should be allotted to each participant nation or group of nations to get their houses in order?

4.8 Sequestration

In spite of all our best intentions to reduce the production of CO_2 by any means available, there remains the recognition that in any foreseeable future the problem will not go away entirely. We are thus left with the issue of possible disposal of CO_2, or at the very least containment. Not all emissions of CO_2 are equally accessible; a large part of the total comes from small, distributed sources that include automobiles and space heaters in homes and work places. These effluents enter the air directly and it is not practical to separate out the offending GHG. Because the gas is only present in the air at the very dilute level of parts per million, there is little hope of removing it by chemical means once the cat is out of the bag. Instead, if it is to be captured at all, it will have to be removed from furnace effluents where it is present in much higher concentrations. While well-established chemical engineering techniques are available for this purpose, the costs of such scrubbing operations are not negligible, and the DOE's Office of Fossil Energy and National Energy Technology Laboratory is supporting basic research aimed at the development of new catalytic processes as well as the application of new materials that can reduce this cost.

But all that only moves the question one step further: what are we going to do with huge quantities of CO_2 once we have captured them? How can we sequester the gas? The study of this question is frequently encapsulated by the acronym *CCS*, which stands for *carbon capture and sequestration*.

The most immediately available method of sequestration is underground containment. There are numerous places where previously used up oil and gas wells have been abandoned. Elsewhere, huge underground salt domes have been discovered, or else large empty volumes are found in association with a water-bearing rock layer, an *aquifer*. The CO_2 could be pumped into these abandoned wells and natural caverns and sealed off from the surroundings. The volumes in these sites are very large, but of course not limitless, and there is some doubt as to whether there is sufficient volume in such repositories to accommodate all the gas that is being released today in addition to that which will be produced later, even though there will be added volume made available as currently active wells become dormant. Added to that possibility is the caveat that some abandoned wells may yet become active when new technology makes further oil or gas recovery again profitable.

These reservations not withstanding, a large coal-fired power plant in West Virginia has been retrofitted to capture CO_2, liquefy the gas, and inject it into a porous dolomite layer 1.5 miles (2.6 km) below the surface. This CCS experiment was scheduled to begin operation in September or October of 2009, with the hope that the CO_2 will be sequestered underground for millenniums.[29] Similar trials are being planned in Europe. One such in the Netherlands is focused on the depleted gas fields that are located more than a mile under the town of Barendrecht,[30] where the Dutch government and Shell Oil want to inject millions of tons of CO_2, starting in 2011. The advocacy group Greenpeace speaks of a slow leak that could follow pipe corrosion if the trapped CO_2 were to dissolve in underground water and form carbonic acid. They would prefer instead to spend the government money on harvesting renewable energy from wind or solar sources. The local townspeople are fearful of a gas leak and object to the plan; it remains to be seen whether or not the government will overrule the town by citing the national interest.

A second type of proposal is for depositing the gas into the deep ocean, where under pressure from the ocean above, the gas would be soluble in the cold water. It would then be expected that it would not bubble to the surface and escape, remaining as dissolved carbonate and bicarbonate salts. In such circumstances, the ocean water in the vicinity would become more acidic, altering ocean chemistry and the biology that depends on that chemical composition. In effect the ocean could become less hospitable to fish, phytoplankton, and other ocean life.[31]

Yet another idea uses again the possibility of burial, but this time in underground deposits of basalt rocks, which are chemical oxides capable of reaction with CO_2 to form solid insoluble carbonates of calcium, magnesium, and/or iron. Such deposits have been found for example under the Pacific Ocean off the coasts of Oregon and Washington State, but there remains some doubt as to whether the necessary chemistry will take place at the needed rates at the local conditions under the sea.

Lastly, it should be noted that controlled photosynthesis has been advanced as a possible mode of removing CO_2 already in the air. The idea is to encourage large-scale growth of plankton either in enclosed containment vessels or in seeded open ocean water.[32] This is not quite a true sequestration, but deserves detailed attention nevertheless. It will be treated further in a later chapter.

In the next chapter, we turn attention to ways in which more direct solar sources could be viable substitutes for fossil fuels. Such sources are generally called *renewable*, although there is on occasion some political disagreement on what exactly may be included under that heading. These matters are also included in the chapter to come.

Notes and References

1. John M. Broder, "E.P.A. Seeks Tighter Rules to Cut Down Air Pollution," *New York Times*, January 8, 2010, p. 1.
2. Joseph E. Aldy and Robert N. Stavins, eds, *Architectures for Agreement: Addressing Global Climate Change in the Post-Kyoto World* (Cambridge, UK: Cambridge University Press, 2007), p. 1.
3. Andrew C. Revkin and Clifford Krauss, "A Cheap, Easy Way to Curb Climate Change: Seal the Gas Leaks," *New York Times*, October 15, 2009, p. 1.
4. A. R. Ravishankara, John S. Daniel, and Robert W. Portmann, "Nitrous Oxide: the Dominant Ozone-Depleting Substance Emitted in the 21st Century," *Science*, Vol. 326, October 2, 2009, p. 123.
5. Monthly mean atmospheric CO_2 at Mauna Loa Observatory, Hawaii, graphed by Hugo Ahlenius, UNEP/GRID-Arendal, http://maps.grida.no/go/graphic/atmospheric-concentrations-of-carbon-dioxide-co2-mauna-loa-or-keeling-curve.
6. IPCC Report: S. Soloman, *et al.*, eds, *Climate Change 2007: the Physical Science Basis. Contribution of Working Group I to the Fourth Assessment Report of the Intergovernmental Panel on Climate Change* Cambridge, UK: Cambridge University Press for the Intergovernmental Panel on Climate Change, 2007), http://ipcc-wg1.ucar.edu/wg1/Report/AR4WG1. Data compiled by the Climatic Research Unit of the University of East Anglia and the Hadley Centre of the UK Meteorological Office. The documentation for this data set is in P. Brohan, J. J. Kennedy, I. Haris, S. F. B. Tett, and P. D. Jones, "Uncertainty Estimates in Regional and Global Observed Temperature Changes: a new

dataset from 1850," *Journal of Geophysical Research*, Vol. 111, 2006, p. D12106.

7. John M. Broder, "Past Decade was Warmest Ever, NASA Finds," *New York Times*, January 22, 2010, p. A8.
8. Morris A. Bender, Thomas R. Knutson, Robert E. Tuleya, *et al.*, "Modeling Impact of Anthropogenic Warming on the Frequency of Intense Atlantic Hurricanes," *Science*, Vol. 327, January 22, 2010, p. 454.
9. Department of Energy, Energy Information Administration (EIA), *Annual Energy Review 2008*, Figure 11.19.
10. John M. Broder and Jonathan Ansfield, "China and U.S. in Cold War-like Negotiations for a Greenhouse Gas Truce," *New York Times*, June 8, 2009, p. A4.
11. Jonathan B. Weiner, *Engaging China on Climate Change*, Resources Issue No. 171, Winter/Spring, 2009, p. 29.
12. S. Soloman *et al.*, eds, *Climate Change 2007*, op. cit.
13. James J. McCarthy, "Reflections On: Our Planet and its Life, Origins, and Futures," *Science*, Vol. 326, December 18, 2009, p. 1646.
14. Elizabeth Kolbert, "The Catastrophist," *The New Yorker*, June 29, 2009, pp. 39–45.
15. Andrew C. Revkin and John M. Broder, "Facing Skeptics, Climate Experts Sure of Peril," *New York Times*, December 7, 2009, p. 1.
16. Eli Kintisch, "Projections of Climate Change Go From Bad to Worse, Scientists Report," *Science*, Vol. 323, March 20, 2009, p. 1546.
17. Stacy C. Jackson, "Parallel Pursuit of Near-Term and Long-Term Climate Mitigation," *Science*, Vol. 326, October 23, 2009, p. 526.
18. Piers Forster, Gabriele Hegerl, Reto Knutti, *et al.*, "Assessing Uncertainty in Climate Simulations," in S. Soloman, *et al.*, eds, *Climate Change 2007: the Physical Science Basis. Contribution of Working Group I to the Fourth Assessment Report of the Intergovernmental Panel on Climate Change* (Cambridge, UK: Cambridge University Press for the Intergovernmental Panel on Climate Change, 2007).
19. Drew T. Shindell, Greg Faluvegi, Dorothy M. Koch, Gavin A. Schmidt, Nadine Unger, and Susanne E. Bauer, "Improving Attribution of Climate Forcing to Emissions," *Science*, Vol. 326, October 30, 2009, p. 716.
20. David D. Parrish and Tong Zhu, "Clean Air for Megacities," *Science*, Vol. 326, October 30, 2009, p. 674.
21. John M. Broder, "Greenhouse Gases Imperil Health, E.P.A. Announces," *New York Times*, December 8, 1009, p. A16.
22. Elisabeth Rosenthal, "Deal Seen Near For Payments to Save Forests," *New York Times*, December 16, 2009, p. 1.
23. John M. Broder, "Obama to Toughen Rules on Emissions and Mileage," *New York Times*, May 19, 2009, p. 1

24. Keith Bradsher, "China's Mileage Mandate," *New York Times*, May 28, 2009, p. B1.
25. Elisabeth Rosenthal, "Buses May Aid Climate Battle in Poor Cities," *New York Times*, July 9, 2009, p. B1.
26. Elisabeth Rosenthal, "In Bolivia, Water and Ice Tell a Story of a Changing Climate," *New York Times*, December 14, 2009, p. 1.
27. Mason Inman, "Hot, Flat, Crowded – and Preparing for the Worst," *Science*, Vol. 326, October 30, 2009, p. 662.
28. W. T. Pfeffer, J. T. Harper, and S. O'Neel, "Kinematic Constraints on Glacier Contributions to 21st-Century Sea-Level Rise," *Science*, Vol. 321, September 5, 2008, p. 1340.
29. Matthew L. Wald, "Refitted to Bury Emissions," *New York Times*, September 22, 2009, p. 1.
30. Aoife White, "Debate over CO_2 Heating Up: an Underground Storage Plan Raises Hackles in the Netherlands," *Philadelphia Inquirer*, November 15, 2009, p. A6.
31. Dalin Shi, Yan Xu, Brian M. Hopkinson, and François M. M. Morel, "Effect of Ocean Acidification on Iron Availability to Marine Phytoplankton," *Science*, February 5, 2010, p. 676.
32. Eli Kintisch, "Rules for Ocean Fertilization Could Repel Companies," *Science*, Vol. 322, November 7, 2008, p. 835.

5

Renewable Energy

Among the several generalizations about energy that were emphasized in Chapter 3 were two points that are especially pertinent in considering renewable sources: first, that each of the various forms of energy can be transformed into an alternative form; and second, that all the forms that we have available came originally from sunlight. That is to say that given some attention to geological time scales, all energy is solar energy.

What then is the distinguishing feature that allows us to label some forms of energy as renewable? It has little to do with the ultimate source and everything to do with the time scale of regeneration. If the time required for

The Challenge of Climate Change: Which Way Now? 1st edition.
By Daniel D. Perlmutter and Robert L. Rothstein.

transformation is measured in hours, days, or months, time duration that is short relative to human lifetimes, we consider the source to be renewable. If on the other hand the time required is long as measured by our experience, we class the supply to be non-renewable. Thus, since formation of the fossil fuels occurred over eons of geological time, they are non-renewable; whereas plant growth via photosynthesis, and changes in wind or tides, are classed as renewable because the transformations of solar radiation into these forms occurs in a matter of hours, days, or months.

This understanding of what is or is not renewable is important as a framework in which to evaluate some claims for renewable status when the moves are political and/or economically motivated. By mid-2009, 28 US states and the District of Columbia had set quotas requiring a percentage of an electricity provider's energy sales or installed capacity to come from renewable resources.[1] To provide economic incentives there are federal tax breaks and extensive new grants and loans available for those installations that fit under the rubric of renewable. Further benefits that go with this designation are *renewable energy credits* which could become salable if proposed national standards become law.

With billions of dollars at stake, lobbyists have been pressing legislators to expand the boundaries of definition and have been able in some states to include as renewable such sources as waste coal and methane from coal mines, and old tires. Other pressures on legislators are to consider the burning of garbage as a renewable process and some want to include nuclear energy under such a title. Recalling that the purpose of moving toward renewable energy is to limit the release of carbon dioxide (CO_2) by reducing our dependence on fossil fuels, this chapter will adopt a narrower view on what energy should be considered renewable. We exclude waste coal and methane from coal mines and nuclear fuels as coming from a finite supply. Garbage and recycled tires can be allowed to the extent that they are in fact shown to be a reliable and ongoing supply. Our focus in the sections that follow will be on the technical and planning aspects of transition from our current status.

One should ask first: where do we stand today? Again we turn to Department of Energy (DOE) data[2] for an answer. As shown pictorially in Figure 5.1, only 7% of all US energy consumption was classed as renewable in 2007. That relatively small segment of the pie can further be subdivided: biomass sources that include wood as well as waste make the largest contribution (53%), hydroelectric power is next (36%), leaving just 10% of all renewable energy attributable to wind and geothermal sources, and only 1% to solar conversion to heat and electricity.

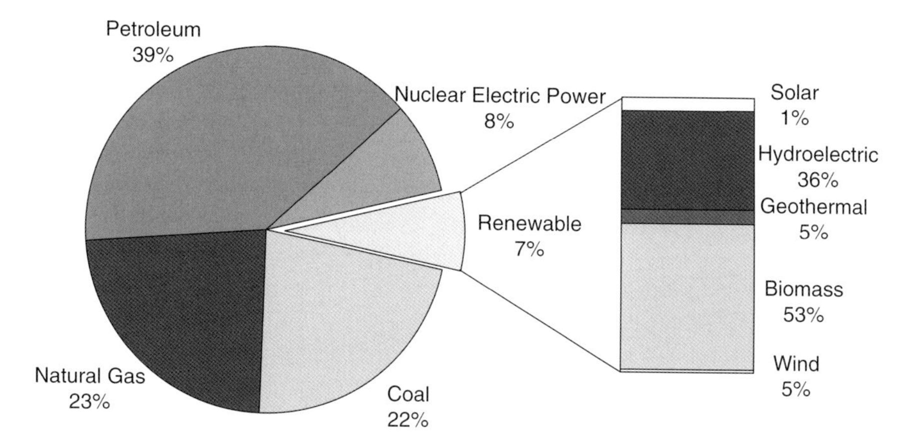

Figure 5.1 US renewable energy consumption, 2007.
Source: US Department of Energy, Energy Information Administration, Office of Coal, Nuclear, Electric and Alternate Fuels

The distribution is somewhat different if attention is shifted to international use.[3] Here hydroelectric power is even more predominant, accounting for about 66% of the total world-wide renewable energy. Combustion of materials of biological origin and crops fermented into alcohol produce some 22%, and wind about 5%. World-wide the conversion of solar radiation into useful energy accounts for 7%, a noticeable increase over the fraction in the US data. A bar graph comparing these magnitudes is shown in Figure 5.2, where geothermal sources are also included for comparison. Geothermal energy is obtained by tapping into steam and hot water heat under the surface, but it is not renewable in a strict sense because the sources are finite reservoirs that can be used up in time. By the same token, nuclear energy that is based on fission is not renewable since uranium is a finite resource. Biological materials are certainly renewable in each growth cycle, although their eventual combustion returns to the atmosphere the CO_2 that was extracted to make possible the growth.

5.1 Hydroelectric Power

The use of water wheels to harvest the energy flowing in streams, rivers, and waterfalls is very old indeed. Such wheels were even of use in the US late into the nineteenth century until they were replaced by steam engines

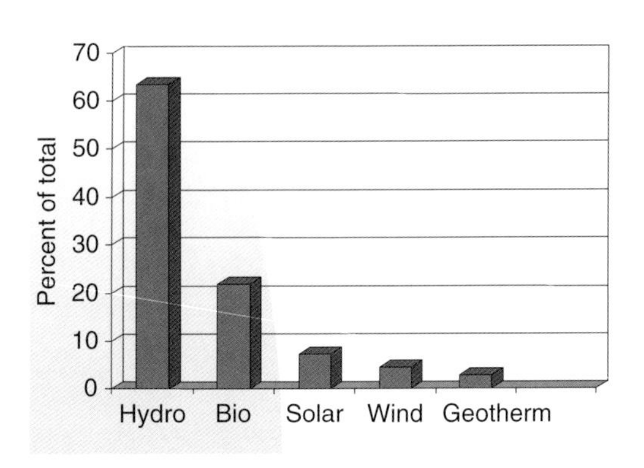

Figure 5.2 World renewable energy, 2009.
Source: US Department of Energy, 2008 Renewable Energy Data Book

and electric motors. Today the water flow is usually held behind large dams and released in a controlled manner to drive turbines linked to generators for electricity. The full capability of a dam (called capacity) is not usually realized because the release of water is governed by added constraints related to supply, maintenance, and irrigation needs. The fraction of the capacity that is actually used is called the *load factor*, expressed as a percent. Table 5.1 lists the capacity and the actual delivery of power for each of the nine nations that are the largest producers of hydroelectric power. The sum of capacity in the table is 542 GW, but this total is not all that could be extracted globally from water power, and dams are being built in many parts of the world to augment the supply. One estimate has the total possible capacity at about 3,000 GW.

The greatest rate of growth in hydro power is found today in China, where the Three-Gorges Project will be the largest in the world. Also in China, the recently completed Xiowan Dam is the world's tallest with water storage capacity equal to all the Southeast Asia reservoirs combined. This dam is one of eight under construction on the upper half of the Mekong River. At the same time Laos has started construction on a series of 23 dams hoped to be completed by 2010 on the lower part of the same river. A recent United Nations (UN) study has expressed concerns that the concomitant changes in

Table 5.1 Hydroelectric capacity by country.

Country	*Capacity GW*	*Load Factor Percent*	*Delivered GW*
China	145	37%	54
Canada	89	59%	53
Brazil	69	56%	39
USA	80	42%	34
Russia	45	42%	19
Norway	28	49%	14
India	34	43%	15
Japan	27	37%	10
France	25	25%	6

river flow will affect biodiversity, damaging the ecosystem which is home to dozens of rare birds and edible marine species.[4]

Whether one looks at the human costs of displaced populations or pays attention to the ecological consequences of dam building, the negative effects must be weighed against the alternatives. With appropriate mitigation of the worst consequences, dams may be worth both these costs if they are compared to the world costs of expanding coal-fired power plants in rapidly growing industrialization. The ideal perfect should not be the enemy of limited improvement.

5.2 Biofuels

There are several major sub-categories of processes that produce products that fall under the title of biofuels. They may use as raw materials woody plant growth from well-established and long used materials such as farm crops, trees grown for this purpose, or forest and field leftovers. New crops such as switchgrass to be grown on otherwise unused fields, or algae harvested in artificial ponds, have also been proposed for this use. Whatever the specific source, the simplest technological treatment is controlled combustion in a furnace; it yields a hot gaseous product (consisting mainly of steam, CO_2, and attendant nitrogen from the air used for combustion) that can be used to drive a turbine and generate electricity in a conventional manner. This route does not, however, yield the desired liquid fuel that is needed for transportation use, nor does it reduce the greenhouse gas (GHG) burden in the atmosphere.

To manufacture a combustible liquid fuel the plant material can be: (i) chemically treated and/or fermented to produce alcohols; or (ii) chemically and/or catalytically converted into hydrocarbons similar to diesel fuel. The first of these alternatives is already in use on a large scale and indications are that it will continue to be important for some years to come. It includes corn and sugar cane grown explicitly for the purpose of conversion into ethanol as a gasoline fuel additive. It also includes a longer range objective: that of converting non-food cellulosic growth into such a fuel. Cellulose is a complex sugar that is a major component of virtually all growing plants, but it is not easily fermented. It must first be broken down by chemical or enzymatic means to produce simple sugars that can then be converted to alcohol.

Biological conversion of crops to fuels has been known since Weizmann's research, by which he was able to produce the butyl alcohol (butanol) and acetone that were badly needed by Great Britain during the years of the First World War. The production method was less than profitable, however, when the exigencies of war subsided and the process was abandoned. More recently, British Petroleum (BP) and Dupont are exploring an updated variation of this process to produce a fuel that they are calling *biobutanol*.[5] They hope to be in large-scale production by 2013, though this may be an overly optimistic projection.

The processes that are already in use, based on fermentation of corn or sugar crops, are very sensitive to national political needs and international ramifications. They provide excellent examples of the intense interaction of technical development, economic benefits to interested parties, and policy decisions. In Brazil, where petroleum is expensive and sugar is cheap, fermentation of sugar cane has been and is today the foundation of their gasoline industry. In the US, ethanol made from corn supplied about 9% of the country's market for liquid fuels in 2009, and the percentage is growing in order to meet the Federal Fuels Standard that mandates an increase from 9 billion gallons in 2008 to 36 billion gallons by 2022.[6] Anticipating the need to meet the new requirements, several large oil companies have entered the field. Sunoco, for example, has purchased an existing factory that is expected to supply 25% of the ethanol that they need to blend into gasoline.[7]

Whether starting from sugar cane or corn syrup, the fermentation process is technically well developed and subsidized in the US by federal grants to farmers and refiners. The benefit of a biofuel is to be sought above all in its reliance on domestic farms to replace imported petroleum, but this objective has been a subject of controversy ever since questions were raised by Pimentel and Patzek.[8] These authors considered the energy used in: (i) raising the crop, (ii) running farm machinery, (iii) irrigating, grinding, and transporting the crop, and (iv) finally fermenting and distilling the ethanol from the water mix; they concluded that more energy is used to produce ethanol in this way than is subsequently released by the ethanol as fuel. Vigorous counter-arguments by representatives of the industry and by reports from the US Department of Agriculture and Argonne National Laboratory reached an opposite conclusion and blamed the discrepancy on Pimentel's use of outmoded data that does not reflect current practice.

The major criticisms of this commitment to ethanol have been two-fold. The first is that the large-scale use of corn to make fuel has caused the price of corn to rise dramatically, thereby hurting all consumers but especially the poor in other parts of the world who depend on US food exports. The second complaint is an outgrowth of the price rise, that it leads farmers throughout the world to convert grasslands and forests into crops. These land clearing practices introduce significant amounts of greenhouse gases into the air, and the changes in landscape remove some of the very active sinks for CO_2 that the world depends on each growing season. One estimate[9] is that the carbon emissions that result from the clearing of tropical forests in places like Brazil, Indonesia, and the Congo now accounts for 17% of all global emissions contributing to climate change.

Searchiger has calculated[10] that burning corn ethanol as fuel produces twice the GHG emissions as gasoline that is alcohol free, if the emissions from land conversion are included in the count. He argues that there is no benefit to the use of biofuels when the full cost to the environment is included in the accounting. This position has been accepted by the California Air Resources Board (CARB),[11] which is charged with putting into practice California's fuel standard, which requires a 10% reduction in GHG emissions from transportation fuel by the year 2020. The federal government is also likely to be drawn into this controversy, since a 2007 law requires the Environmental Protection Agency (EPA) to calculate "life cycle greenhouse gas emissions" for renewable fuels.

To circumvent some of the difficulties associated with the use of food crops to make liquid fuels, there is research in progress that would use non-food crops (i.e. cellulosic feedstocks) as raw materials for ethanol manufacture or to form other fuel components. The intention is to generate these feedstocks from perennial crops grown specifically for this purpose, sited on marginal lands to prevent competition with food production.[12] Steps in this direction are supported by the 2007 US legislative mandate for 16 billion gallons of cellulosic ethanol by 2022, as well as by the European Union's directive that 10% of all transport fuel in Europe should come from renewable resources by 2020. Congressional passage of the 2008 Farm Bill further addressed biofuels and provided subsidies for the production of cellulosic ethanol and biodiesel fuel to the extent of $1 per gallon for refiners and $45 per ton of biomass for growers. This federal tax credit was allowed to expire at the beginning of 2010, making these processes much less attractive economically. In addition much of the overseas market dried up when the European Union put in place tariffs on all biofuels.

The second of the sub-categories of processes that produce liquid fuels, those that focus on hydrocarbons as their goals, are as yet untested on any scale beyond the laboratory. The processes that are currently being studied as candidates for large-scale expansion[13] are based on a variety of chemical steps that convert the oils extracted from plants or algae into compounds similar to those found in petroleum products. These could be used as gasoline, diesel fuel, or jet fuel, depending on the detailed conditions of the conversion steps. Plans for the works in progress, with target dates as early as 2011 to 2016, call for factories with capacities to produce 100 million gallons of fuel annually at prices that would be competitive with petroleum at $60 per barrel. Exxon-Mobil, for example, has announced an intention to invest $900 million over a 5-year period to develop a process to produce refined liquid fuels from algae.

Yet another variation on this theme is based on a high-temperature chemical gasification of the raw material to produce an intermediate gas called *syngas* (short for synthesis gas). In turn, the so-called syngas (a mixture of hydrogen and carbon monoxide) is reacted further to form the desired hydrocarbons catalytically. One can also think of this process as a technique for storing, first in syngas form and then in hydrocarbon fuel, part of the chemical energy that was located in the plant raw material by photosynthesis. With this in mind, we will discuss the features of this process further in Chapter 6 on energy storage.

Whatever the details of processing, it should be added in assessing the potential benefits of using so-called *biomass* as a raw material that any biomass is a bulky solid with quite high water content. The relative costs of shipping and drying mitigate for processing plants that will be close to the biomass source and thus tends to constrain the size (and efficiency of scale) of the manufacturing facility.

What may emerge to be the most telling argument in the debate around biofuel alternatives is presented very sharply by Cambell, Lobell, and Field.[14]

They addressed directly the two current alternatives for biomass use: (i) conversion to ethanol to power internal combustion vehicles; or (ii) conversion to electricity to power battery electric vehicles. Their life cycle assessments compared the transportation distances and greenhouse gas reductions that would be achieved from commitment of land area to one of these choices, accounting for energy needed to grow the feedstock and convert it to either electricity or ethanol, as well as the energy needed to manufacture and dispose of vehicles. They found that "one can travel farther on biomass grown on a hectare of land when it is converted to electricity than when it is converted to ethanol." The results of their work also show that GHG emissions are less when the electricity route is followed in preference to the alcohol route, even when land use impacts are left out of the calculations. Other indirect advantages of the electric route may come from the easier connection with other renewable sources such as solar and wind power, or even greater benefit from centralization if CO_2 sequestration ever becomes accessible.

5.3 Wind Power

As with water power, the use of wind to drive windmills has a long established history. What is new is the advanced machine technology linked to electricity generation that offers more efficient collection of the energy. The energy collected by a windmill is very sensitive to wind velocity (it varies with the square of the velocity; i.e., a velocity reduction in half has the effect of cutting the energy collection to a quarter). As a result it is essential to place windmills at sites of frequent high winds. As illustrated in Figure 5.3 the preferred locations in the US are in the center, north, and west of the country, as well as along the Atlantic and Pacific coastlines.[15] Wind velocities also vary with time of day and season of the year, typically peaking in summer and falling to a low in mid-winter. As a result, wind farms only deliver a fraction of their rated capacity, typically about a third. This means in effect that one cannot depend on wind energy alone, but must plan on integration with other more dependable sources.

As of the year 2007 the installed windfarms in the US had a cumulative capacity of about 18 GW of power. This was increased to 25 GW in 2008, and grew further to 35 GW by the end of 2009, aided by federal tax credits and investment incentives as well as state laws that mandate that some fraction of local power come from renewable sources.[16] Plans are in place for

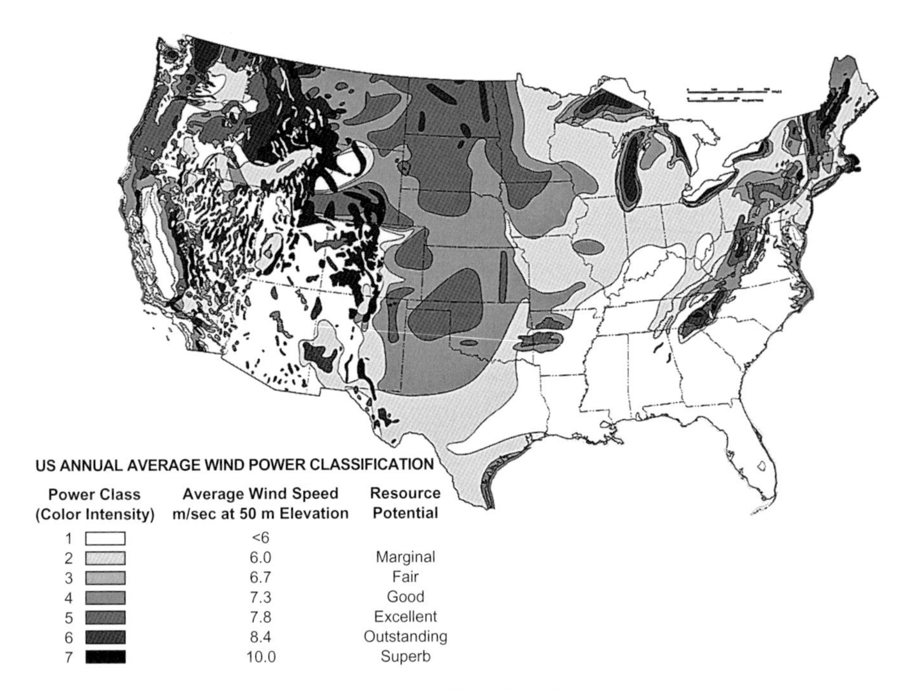

Figure 5.3 US wind power resources at 50 m elevations.
Source: US Department of Energy, National Renewable Energy Laboratory, Wind Energy Resource Atlas of the US

sizable expansions both on land and in offshore locations. The US pattern of growth in generating capacity from this source is shown in Figure 5.4 for the years 1981–2008.[17]

In the UK about 2.5 GW of wind power currently exist, an additional 8.5 GW have been approved but are not yet built, and further increase of 22 GW have been proposed to be developed by installing 7,000 new wind turbines. If all goes according to this plan, the total capacity will be 33 GW, a very significant addition to the 75 GW which currently exist in the UK from coal, gas, nuclear, and hydro power. The goal set by the European Commission is to obtain 20% of its electricity from renewable sources by 2020; Denmark has already reached that target. From 2006 on, China has been making huge investments in green technologies, aiming to increase massively energy generation from wind, solar, and other renewables by 2020.[18]

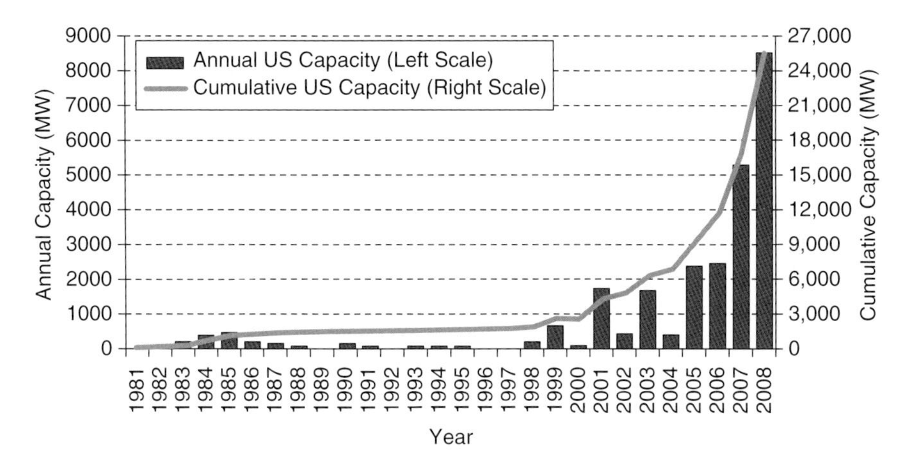

Figure 5.4 Growth of US wind electricity generating capacity.
Source: US Department of Energy, Energy Information Administration, Office of Coal, Nuclear, Electric and Alternate Fuels

5.4 Power from Tides and Waves

The harvesting of electric power from ocean waves was seriously proposed[19] as early as 1974, but the outcome can be easily summarized: the source has always been available and never been used because the investment has always been considered to be too large to justify the savings in energy. Today the picture is changing as energy costs rise and the threat of global warming is ever present. Scruggs and Jacob[20] reported on a finding of the US Electric Power Research Institute (EPRI) that the potential ocean wave energy in the US is comparable to energy currently generated by conventional hydro power. They also cited an estimate by the Carbon Trust in the United Kingdom to the effect that their economically viable offshore resource amounted to about 14% of their current national demand.

To find suitable locations at which to extract energy from the sea, one seeks low wave frequencies and high amplitude waves. This means that deeper waters are preferred to shallow beaches and western shores are typically better than those that are east facing. Waves form in deep water and their energy dissipates as they approach a shore; at depths less than 20 m the energy is less than one-third of that carried by the wave in deep water. As of 2009 existing installations are off the coasts in Portugal (rated at 2.25 MW), Spain (1.4 MW), and Oregon (2 MW). There is as yet no agreement on the best design, and the various test units are small and use different technologies and geometric arrangements. They are typically tuned to work best with a given wave frequency, but real waves exhibit random behavior over a range of frequencies, and statistical expectations and close control are needed to optimize an entire system.

Although tides and waves both present forms of water power, they exhibit several important differences. For one, tidal power is the only form of energy that derives directly from the motion of the moon about the earth and the movement of planet earth as it orbits the sun. These relative motions produce gravitational shifts that periodically change the water levels that we recognize as high and low tides. The size of the tidal fluctuations at any given location depends on the changing positions of the moon and sun relative to our planet, but also on the shape of the coastline, the slope of the continental shelf, and the shape of the sea floor.

Because a tidal energy generator uses this phenomenon to generate energy, tidal power is highly predictable and practically inexhaustible. It does, however, vary sharply with geography and the location for tidal energy generation should be chosen with water level height or tidal current velocity in mind, typically at the mouths of bays and rivers or between land masses. One type of arrangement, called a *barrage*, is in essence a dam built across a tidal estuary; it holds back water to store it for its potential energy and releases the stored supply through turbines when the energy is needed. A major tidal power station of 240 MW capacity, the largest barrage in the world, has been operating in La Rance, France, since 1966. In Britain a huge 2 GW capacity barrage is planned to utilize by 2028 the 15 meter tide on the Severn River. The capital cost is estimated to be $29 billion, expected to offer long-term returns with minimum maintenance cost and a projected savings of over 19 million ton of coal each year. However, blocking the mouth of an estuary can have significant ecological effects similar to those that occur at any large dam, changing local marine life and vegetation. An alternative approach harvests the kinetic energy in the moving water by placing turbines directly into a moving stream. One such, for example, with a capacity of 1.2 MW and billed as the world's largest tidal turbine, was installed in 2008 in Northern Ireland's Strangford Lough.

The distinction made between tides and waves should be emphasized again with regard to their dependability, because whereas tides are predictable over short and long time scales, wave action depends to a great degree on driving winds. As a consequence, energy recovered from wave action will suffer from the same irregularities as that recovered via wind turbines.

5.5 Direct Use of Solar Energy

The point has been made and emphasized before that all renewable energies either directly or indirectly use solar radiation as an original source. Thus wind is created by the sun unevenly heating the air. Rain and snow, which flows to all rivers, form when the air is cooled sufficiently to cause previously evaporated water to condense and precipitate. But when we speak of solar energy, we mean direct use of the sun's radiation as opposed to indirect use via wind or waves.

The direct use of sunlight is by one of two transformations: (i) the sun's radiation can be used to generate heat in whatever fluid is designated to be the carrier; or (ii) radiation can be absorbed in photovoltaic (abbreviated PV) cells that directly convert the energy to electricity. The direct conversion to electricity is accomplished by specially designed silicon cells that absorb light and put out an electric current, but cheaper substitutes for silicon based on cadmium telluride thin films are also being worked on, as are methods that use reflectors to focus more sunlight on a smaller collector area.

In residential use the solar energy is usually collected by heating a circulating fluid. The carrier is hot water that is then passed through piping for internal space heating or bathing. For industrial applications that can efficiently use high-temperature heat, the working fluid can be a molten salt that gives up its energy to create steam for power generation. In this case focusing mirrors have been suggested in order to cover a wider area for collection of sunlight, which is reflected in the direction of a central receiver.

To encourage research and development in this arena governments have provided incentive policies for PV cells. These include tax benefits that serve to refund part of the investments needed to build and install systems, as well as constraints on utilities that require them to buy PV electricity from producers. In response the largest utility in New Jersey announced in 2009 a plan to install 80 MW of PV collectors over a 4-year period.[21] In Germany, California, and Florida governments have introduced a so-called "feed-in tariff" which sets the price of the electricity sold to the utilities above the going market price.[22] It is estimated that by the end of 2008 about 90% of the PV-generating capacity was tied into the utility electricity grid. The political purpose of these subsidies is to promote greater national energy independence and facilitate start-up and growth in an industry not yet ready to compete until it reaches the necessary economies of scale. The immediate effect of scale can be seen by recalling that PV cost was as high as $25 per watt in 1979. Current costs of silicon flat panel modules are priced at about $5 per watt, increasing to \$6.4 per watt with included installation charges in 2009. To cite some typical figures,[23] a new 80 kW rooftop solar array installed in a Philadelphia factory is reported to cost \$536,000, corresponding to \$6.7 per watt, but after federal and state grants and rebates, along with depreciation tax credits, the net cost (including a new roof) is reduced to \$195,000, giving an installed cost of \$2.4 per watt. A comparison study by Greg Nemet[24] concluded that 43% of the drop in cost in the 22 years since 1979 was attributable to economies of scale. He associated another 35% of the reduction to progress in research and development.

One scheme that can improve collection efficiency of solar cells uses reflectors to pick up radiation over a larger area. This reduces cost, because the reflecting collector is less expensive than an equivalent area of solar cells. Whereas silicon flat panel modules with included installation charges are priced at about \$6.4 per watt, the system with reflecting collectors is reduced in price to about \$3 per watt. Estimates by those within the industry[25] are understandably very optimistic, expecting that as manufacturing costs fall, PV plants will be able to compete with more standard electricity generation by 2014, even without the current US federal incentive of a 30% investment tax credit. Nevertheless, to put these estimates in context it should be noted that the total accumulated world-wide capacity of PV production was about 15 GW in 2009, half of which was in Germany and around 10% in the US. This means that as of that date the entire PV megawattage amounts to less than 0.5% of the world installed electricity generating capacity.

As is so often the case when new high-tech devices enter the market, the cost of PV conversion is still too high to compete with conventional alternatives except for special situations where sites are too far removed from a supply grid, or in areas with abundant sun and high costs for electricity such as in parts of California, Japan, or Hawaii. On the other hand, the cost of solar thermal power is competitive with clean coal (without sequestration) and is less expensive than nuclear power for small-scale installations. Comparison numbers from a recent National Research Council report[26] on each of these alternatives for generating electricity are given in Table 5.2, where the listings are energy costs per kW used for an hour (cents/kwh). Given these numbers, it is not surprising that coal burning accounts for 49% of US electricity generation, natural gas for 21%, and nuclear power for 20%.

Ignoring for the moment all questions of economics and costs, there is yet another consideration that puts power from PV sources into sharper perspec-

Table 5.2 Costs of US electricity in 2005.

Source	*Cents/kwh*
Conventional Coal	4
"Clean Coal" (without Sequestration)	7
Nuclear	11
Solar Thermal	8
Solar Photovoltaic	>24

tive as a possible replacement for current sources of energy for electricity generation. We should ask: how much surface area would need to covered by silicon cells to obtain a useful amount of electric energy from the sun via PV? Allowing for reflection, cloud cover, weather, and the many vagaries during the year, the energy that reaches the surface of our world averages about 200 W/m^2 (watts per square meter). We might expect current PV conversion to be 10% efficient and supply 20 W/m^2 as electricity.

One recent estimate[27] projected that by 2050 the US could be using existing PV technology for 69% of total electricity consumption, i.e., 700,000 MW. A simple ratio of these numbers leads to the area requirement of 35 trillion square meters or approximately 14,000 square miles. Allowing for space between the PV modules and making room for auxiliary equipment, service vehicles, and adjacent connectors and carriers of electricity, one might reasonably double the area required to 28,000 square miles, an area equivalent, for example, to 23% of the total area of the state of New Mexico. This area requirement is certainly not a trivial consideration, but the overall picture might improve if the collectors could be sited exclusively in desert areas where the incident radiation could be expected to yield more than 20 W/m^2, or if the collectors' efficiency for energy conversion were to increase significantly. Given such improvements the plans would still need to take into consideration the costs of power transmission over long distances to the consumers.

If instead we focus attention on a single home with say 100 m^2 of useful roof area, the same 20 W/m^2 could supply 2 kW of power, enough to meet the monthly electricity demand of the average US household; however, this monthly estimate is an average over day and night, rain and shine, and would not be sufficient at times of peak demand when supplemental sources would have to be called upon. One way or another each of the renewable energy sources are intermittent. This will lead us into Chapter 6, where attention is to be paid to options for storage of energy.

5.6 Nuclear Energy

It was noted above that nuclear energy is not fully renewable since the uranium supply is finite, but whatever its proper category, nuclear power has some obvious advantages among the possible ways to generate electricity. Since they do not burn hydrocarbon fuels, operation of these plants does not produce any CO_2. Above all, nuclear power is a known technology. In France some 80% of all electricity generation comes from nuclear plants, and in the US more than a hundred commercial reactors generate almost 20% of the country's electric power right now. In spite of that not a single new plant has been ordered and built in the US in over three decades. The lack of any new moves in this regard was a response to increased costs of construction as well as heightened fears of radiation hazards following the 1979 Three Mile Island accident in the US and the 1986 explosion in Chernobyl, Ukraine. In addition there has been ongoing discomfort with the current above-ground storage of nuclear wastes, currently being held at multiple sites scattered all over the US. A proposed unification to a single underground site in Yucca Mountain in Nevada has been delayed for decades by geological uncertainties, changing performance expectations, and local political opposition.[28] Now, after some $10 billion has been spent on risk assessment studies, it appears from plans for the 2010 federal budget[29] that a new strategy for nuclear waste disposal will replace the earlier idea to use the Yucca Mountain site.

Today, with the rising public concern over global warming, we may be entering a period of nuclear revival.[30] In 2008 the Nuclear Regulatory Commission had applications for permission to build 34 new plants, and Congress has provided loan guarantees and insurance against regulatory delays. The implementation of such guarantees is yet another source of controversy, however, because major delays and cost overruns have in the past been common in reactor construction.[31] With government guarantees any private sector losses would have to be covered from the US Treasury. In spite of this easing of impediments, it must be recognized that new construction of nuclear plants takes many years from design to construction to completion.

As a remedy to fossil fuel use, this avenue is not an available short-term fix and should properly be thought of as an intermediate-term target.

Expansion of nuclear capacity is also being planned abroad. China had announced plans 3 years ago to move from a capacity of 9 GW to 40 GW by 2020; they have now stepped up their expected development in this area to aim for 70 GW capacity by 2020 and 400 GW by the year 2050. If the Chinese meet their targets for 2020, they estimate that nuclear power will still only provide about 10% of their electricity needs. Aware of the hazards associated with this industry, China has requested experts from the International Atomic Energy Agency for staffing and training help.[32]

5.7 Geothermal Energy

Below the thin layer of our planet on which humans live, the world is hot and has been so from the time of its origin to the present. To the extent that we have access to this huge thermal reservoir it can provide a sustainable source of energy, but in most locations it is cut off from us by an insulating crust. In places where the crust is thin or broken, generally near tectonic plate boundaries, we can and do harvest energy by passing steam or hot water through the underground reservoirs and using the energy to drive electrical generators or to provide thermal space heating. As of 2007, the world-wide electricity generating capacity from this source was about 10 GW, and the estimated thermal heating amounted to an additional 28 GW, both quite small in comparison to conventional fossil fuel sources.

To access deeper pockets of heat and locations elsewhere away from tectonic boundaries it is necessary to drill through layers of rock and earth, usually to depths of several miles below the surface. Such techniques have been developed for petroleum production, but the costs of drilling have until now discouraged this endeavor. The investment in so-called *enhanced geothermal systems* will become more attractive as other fuels become more expensive and the reduction of CO_2 release more pressing. In fact concentrations of CO_2 are present in some of the hot gases under the crust, but they can be recycled back into the same holes from which they were released. In one variation of this approach, pressurized water is injected deep in the ground with the object of cracking the rock that is trapping underground heat; however this process has been associated with local earthquakes. A 2006 project in Switzerland had to be stopped when many thousands of seismic events were recorded and felt during 6 days of water injection, and the project was shut down permanently in 2009 in response to the determinations in a Swiss government study.[33] A similar 2009 project in California[34] has raised earthquake fears among residents because it is designed to drill over 2 miles (3.2 km) below the surface. Elsewhere, in New Zealand and Germany, geothermal projects have caused subsidence of the bordering lands.

In spite of these cautionary tales, the advantages of geothermal energy are very significant: the supply is virtually inexhaustible, it is available at all times of the day and all seasons of the year, and the cost of production is low except for the initial drilling investment. And above all, this source promises to reduce the overall greenhouse gas burden, and there is still hope that a practical operation might work in a less populated area.

5.8 Indirect Emissions and Hidden Costs

The full effects of any change are not always apparent. When the California Air Resources Board had the task of evaluating the addition of alcohol to gasoline, for example, their finding was determined not merely by the immediate carbon release from the combustion of the fuel, but crucially by a secondary effect: the emissions that came from the land conversion that followed the economic pressures created by the new demand for crops to be fermented. The realistic assessment called for an overall view in context, recognition of the subsequent and indirect effects on the overall carbon balance. A similar evaluation is needed for each and every proposed step toward climate control.

A point of particular sensitivity in this regard has been raised in relation to nuclear power.[35] At first glance, it appears that a nuclear reactor does not produce CO_2 at all, that the fission of uranium produces only "green" energy in the form of steam that is used to drive turbines and thus generate electricity. A closer examination of the full process must recognize, however, that the production of uranium and/or plutonium is not without energy input. First, the uranium ore must be dug from its source. Then the uranium 235 isotope must be concentrated, originally via gaseous diffusion and more recently by using high-speed centrifuges. Finally, the uranium must be transformed into oxide pellets and fuel rods suitable for insertion into the reactor core. Each of these steps demands energy input, in some cases using fossil fuels, in other steps using electricity generated by combustion of fossil fuels. The full effect is surely to release CO_2 into the atmosphere, but the extent of this release is not generally documented, perhaps because of its association with weapon uses of uranium, and as a consequence it is not easy to estimate. In the book cited above, Helen Caldicott indicated that the creation of nuclear electricity produces one-third as much CO_2 as a similar-sized conventional plant that burns natural gas. She cautioned, however, that a still greater ratio of fossil fuels will be needed in the future as the quality of available uranium ores decreases.

Yet another uncertainty in this arena arises from the possible recovery of uranium by reprocessing spent fuel. To date this has not been part of the US nuclear plant protocol, although it is done in other nations, notably France. Such reprocessing will become increasingly attractive if and when the cost of newly refined uranium rises, but of course this mode of chemical recovery will also require an investment of energy. Furthermore, as with any natural resource the cost depends on ease of availability, and conversely, additional amounts become economic to recover as the price goes up. Data on reserves of uranium oxide have been published by the DOE[36] and are shown in Figure 5.5 as proved, estimated, and speculative reserves for each of three costs per pound. As points of reference, it should be noted that the price of uranium oxide has varied since 1981 from a low of $10 to a high of $43 per pound.

The costs of implementing change are not always directly expressed in terms of construction dollars or operating dollars. Any change, for example, that calls for increased demand in irrigation for agriculture or in cooling water for power generation will have to deal with water supply issues. Depending on location, a large growth in demand could trigger water shortages, costs that are great even if not directly expressed in simple monetary terms. Robert F. Service has reported[37] on studies that say 98 gallons of irrigation water are required on average to produce each gallon of alcohol via the corn fermentation process. This translates into an increased irrigation need of at least 2 billion gallons per day if US farms are to produce enough crops to meet the congressionally mandated production of alcohol and other advanced biofuels. To put these figures in perspective, they may be compared to the water needs of more conventional power generation: for the same energy production, corn ethanol irrigation requires at least 30 times as much water as a power plant burning natural gas.

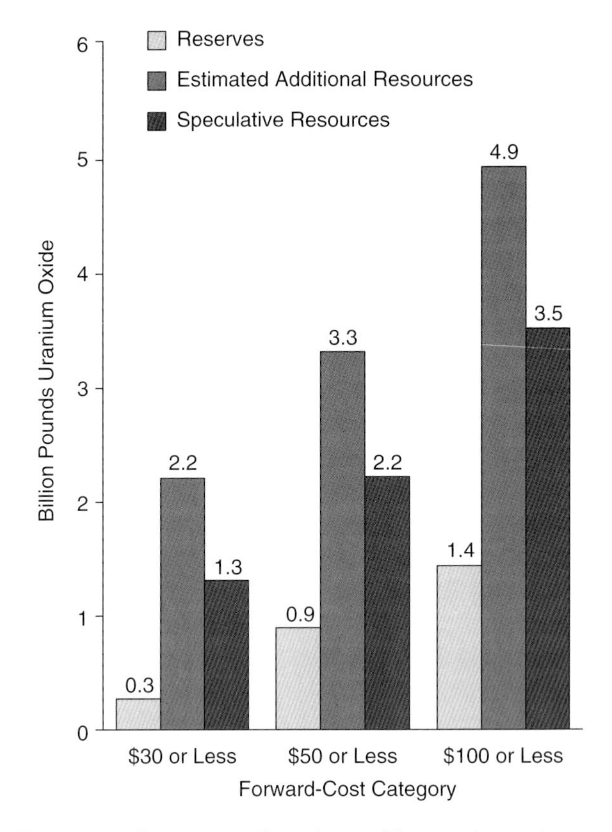

Figure 5.5 Reserves and resources of uranium oxide at various prices.

Notes and References

1. Felicity Barringer, "With Billions at Stake, Trying to Expand the Meaning of 'Renewable Energy'," *New York Times,* May 25, 2009, p. A8.
2. US Department of Energy, Energy Information Administration (EIA), *Annual Energy Review 2008*, published June, 2009, Figure 10.1.
3. US Department of Energy, *2008 Renewable Energy Data Book*, published July, 2009.
4. Michael Casey, "Dams Pose a Danger to Mekong Waterway," *Philadelphia Inquirer, May* 24, 2009.
5. Glenn Hess, "BP and DuPont Plan 'Biobutanol'," *Chemical and Engineering News*, Vol. 84, No. 26, June 26, 2006, p. 9; also at http://

green.blogs.nytimes.com/2009/07/16/biobutanol-creeps-toward-the-market/?scp=1-b&sq=biobutanol+BP&st=nyt.

6. Clifford Krauss, "Big Oil Warms to Ethanol," *New York Times, May* 27, 2009, p. B1.
7. Harold Brubaker, "Sunoco Wins Ethanol-plant Bid," *Philadelphia Inquirer, May* 25, 2009, p. C1.
8. David Pimentel and Tad W. Patzek, *National Resources Reseach*, Vol. 14, No. 1 (2005), pp. 65–76.
9. Thomas L. Friedman, "Trucks, Trains and Trees," *New York Times, Novembe*r 11, 2009, p. A27.
10. Timothy Searchiger, Ralph Heimlich, R. A. Houghton, et al., "Use of U.S. Croplands for Biofuels Increases Greenhouse Gases through Emissions from Land-use Change," *Science*, Vol. 319, February 29, 2008, p. 1238.
11. Dan Charles, "Corn-based Ethanol Flunks Key Test," *Science*, Vol. 324, May 1, 2009, p. 587.
12. G. Philip Robertson, Virginia H. Dale, Otto C. Doering, et al., "Sustainable Biofuels Redux," *Science*, Vol. 322, October 3, 2008, p. 49.
13. John R. Regalbuto, "Cullulosic Biofuels – Got Gasoline?" *Science*, Vol. 325, August 14, 2009, p. 822.
14. J. E. Campbell, D. B. Lobell, and C. B. Field, "Greater Transportation Energy and GHG Offsets from Bioelectricity than Ethanol," *Science*, Vol. 324, May 22, 2009, p. 1055.
15. US Department of Energy, *Renewable Energy, Wind Powering America*, http://www.windpoweringamerica.gov/wind_maps.asp.
16. Jad Mouawad, "Wind Power Grows 39% for the Year," *New York Times, January* 26, 2010, p. B1.
17. US Department of Energy, Energy Information Administration (EIA), Office of Coal, Nuclear, Electric and Alternate Fuels, *Wind Energy Update April 2010*, http://www.windpoweringamerica.gov/pdfs/wpa_update.pdf-8031.5KB.
18. Robert Kennedy, Jr., "The New (Green) Arms Race," Outreach (Copenhagen: Stakeholders Forum), December 7, 2009; also see Martin Jacques, *When China Rules the World: the End of the Western World and the Birth of a New Global Order* (New York: Penguin Press, 2009).
19. S. H. Salter, "Wave Power," *Nature*, Vol. 249, 1974, p. 720.
20. Jeff Scruggs and Paul Jacob, "Harvesting Ocean Wave Energy," *Science*, Vol. 323, February 27, 2009, p. 1176.
21. Andrew Maykuth, "PSE&G Plan Takes Solar Energy Public," *Philadelphia Inquirer, July 30*, 2009, p. 1.
22. Michael Scott Moore, "Germany's Fine Failure," *Miller-McCune,* Vol. 2, No. 4, p. 15.
23. Diane Mastrull, "A Solar Investment," *Philadelphia Inquirer*, November 15, 2009, p. C1.

24. Greg Nemet, *IIASA Annual Report for 2008*, http://www.iiasa.ac.at.
25. K. Zweibel, J. Mason, and V. Fthenakis, "By 2050 Solar Power could End US Dependence on Foreign Oil and Slash Greenhouse Gas Emissions," *Scientific American,* Vol. 298, No. 1, 2008, pp. 64–73.
26. National Research Council, *Critique of the Sargent and Lundy Assessment of Cost and Performance. Forecasts for Concentrating Solar Power* (Washington, DC: National Academies Press, 2002).
27. Richard M. Swanson, "Photovoltaics Power Up," *Science*, Vol. 324, May 15, 2009, p. 891.
28. Rodney C. Ewing and Frank N. von Hippel, "Nuclear Waste Management in the United States –Starting Over," *Science*, Vol. 325, July 10, 2009.
29. Office of Management and Budget, *A New Era of Responsibilities: Renewing America's Promise*, (Washington, DC: Government Printing Office, 2009), pp. 63–5, http://www.whitehouse.gov/omb/assets/fy2010_new_era/Department_of_Energy.pdf.
30. Matthew L. Wald, "Nuclear Power may be in Early Stages of a Renewal," *New York Times, October* 23, 2009, p. B3.
31. Matthew L. Wald, "Loan Program may Stir Dormant Nuclear Industry," *New York Times*, December 23, 2009, p. B1.
32. Keith Bradsher, "China, Rushing into Reactors, Stirs Concern," *New York Times*, December 10, 2009, p. 1.
33. James Glanz, "Quake Threat Leads Swiss to Close Geothermal Project," *New York Times*, December 11, 2009, p. A12.
34. James Glanz, "In Bedrock, Clean Energy and Quake Fears," *New York Times*, June 24, 2009, p. 1.
35. Helen Caldicott, *Nuclear Power is Not the Answer* (New York: The New Press, 2006).
36. US Department of Energy, Energy Information Administration (EIA), *Annual Energy Review 2008*, Figure 4.13.
37. Robert F. Service, "Another Biofuels Drawback: the Demand for Irrigation," *Science*, Vol. 326, October 23, 2009, p. 516.

6

Energy Storage

In our examination of the possible sources of renewable energy there emerged one troubling observation that was characteristic of virtually all the alternatives, with the exception of tidal energy: the sources were intermittent. Because half the world is in darkness half of each day, there is no solar energy to be harvested then even if the sky is clear. The interruption of input will also be appreciable during daylight hours on rainy or cloudy days. Similarly, power from wind will be reduced or die completely whenever the wind velocity falls sharply. One approach to a remedy for these failings is to transport energy over large distances from places where it is available to those parts of the globe where the energy is needed, but this idea presupposes that the costs of transportation are not prohibitive. In some locations the investment

The Challenge of Climate Change: Which Way Now? 1st edition.
By Daniel D. Perlmutter and Robert L. Rothstein.

needed for the installation of towers and power lines to carry electricity can approach $5 million per mile, a significant cost in assessing practical windmills or solar devices that must be located far from population centers where sufficient sun and wind are available.

There are yet other limitations that arise from adopting increased dependence on renewable sources. First, usable energy is called for at specific times, but not always when it is available. This peak usage must be supplied from supplementary sources that can be tapped at short notice. Second, it should be noted that much of our energy use is for cars, trains, planes, and ships. For the foreseeable future, these outlets will require portable fuels, usually in liquid form and not immediately supplied by renewable sources. All these considerations point in one direction: there is a pressing need for holding energy in storage, for keeping it in some temporary hold that would circumvent the limitations of intermittent source, peak demand, and portability requirements. Moreover, if the energy transferred into storage can be transmitted as electric power during off-peak times, the reduced load on power lines during peak times will allow for lesser transmission capacity and corresponding savings in investment costs.

In this chapter we will consider a variety of energy reservoirs. Several of the storage alternatives are in chemical forms; that is, whatever energy we wish to store will be used to form chemicals from which energy can later be extracted. The recovered energy might emerge in the form of electricity as from a battery or fuel cell, or it could be obtained as heat from combustion of a liquid or gaseous fuel. For some purposes it is advantageous to store energy in a potential form, either by creating a large reservoir of water at an elevated location, or by storing the energy in the form of compressed air. In both of these cases the deposited energy would be withdrawn by driving a turbine or other immediately useful machines. Further in this chapter, we will consider storage in the form of thermal energy in a heated fluid. The fluid could be any carrier of heat, but the commonly found suggestions are to heat water or molten salt. Finally, we will complete the survey of this chapter by adding reference to one more alternative device, a flywheel that could be used to store energy in the form of kinetic energy.

6.1 Batteries and Fuel Cells

Batteries are familiar in everyday use to start cars and to power flashlights, toys, watches, computers, telephones, and myriad other household and transportation applications. They are chemical devices that store electric energy, but ordinarily we only think of their storage limits when they run down. On a small scale, as for watches and flashlights, it is common to dispose of a battery and replace it when it has lost its charge. This works because these small batteries sometimes serve us well for years, a period during which the rate of withdrawal of energy is small compared to their fully charged capacity. On the other hand many batteries have to be recharged periodically in order to provide practical service, as is the case, for example, for cellphones, portable radios, and laptop computers, as well as for the well-known lead-acid automobile battery, which is automatically recharged as one drives.

All batteries have in common two poles, a positive anode and a negative cathode. They differ in that they use different chemical entities for their anode and cathode, as well as for the conductor that serves to transfer charges between the poles. Each combination has its own characteristics and is able to provide advantages in particular applications. To illustrate this point it is instructive to consider several candidates that have been studied as batteries for electric cars and automobile hybrids. In addition to the costs of materials and fabrication for a given weight and volume of batteries that can be carried, there are three other primary requirements: an acceptable range of travel between recharges of the battery, the ability to adequately accelerate the vehicle when needed, and an acceptable number of cycles of charging, discharging, and re-charging before the battery is useless. Translating these expectations into design specifications will demand: (i) a range of at least 300 miles (480 km) between charges; (ii) an installed battery capable of generating about 150 watts per kilogram of weight (W/kg); and (iii) a number of cycles greater than 1,000 that can give a battery life of at least 5 years.

These characteristics are shown in Table 6.1 for five battery types that have been looked at as possible automotive candidates. The familiar standard lead-acid battery is a good starting point. It is acceptable with respect to acceleration and one might perhaps compromise on the number of charging cycles, but it is very heavy for its power delivery and would not be acceptable in an all-electric car because it would need to be recharged after only some 130 miles (210 km). The nickel-cadmium combination tolerates a greater number of cycles, but it also fails to offer the range that is desired. The sodium-sulfur combination could supply the range that is needed, but the materials are fire hazards if they are exposed to the air by an accident. The nickel-metal hydride and the lithium-ion batteries both allow a large number of cycles and an augmented potential for acceleration, making them good candidates for the

Table 6.1 Candidate batteries for automotive use.

Type of Battery	*Estimated Miles/ Charge*	*Relative Acceleration (in W/kg)*	*Number of Recharge Cycles*
Lead-Acid	130	165	650
Nickel-Cadmium	185	150	1,500
Sodium-Sulfur	500	200	600
Nickel-Metal Hydride (Prius before 2009)	300	600	1,000
Lithium-Ion (for Prius after 2010)	500	600	1,200

requirements of a hybrid car that uses its gasoline engine to periodically recharge its battery.

Equally important in choosing a battery are its weight and volume requirements. The nickel-metal hydride combination is much smaller in volume than either the lead-acid or nickel-cadmium battery, and the lithium-ion battery is both smaller in size and lighter in weight than the competition. In fact the nickel-metal hydride was the commercial choice in the Toyota Prius until at least the 2009 models, and General Motors and Toyota have both announced plans to produce a small number of commercial vehicles after 2010 that will be based on the lithium-ion battery.

A variation on the theme of hybrid automotive power is a type of electric car called a plug-in hybrid. The intention here is to run the vehicle as much as possible from the battery, using the gasoline engine as the power source only if extra power is called for or when recharging is not practical. Ordinarily, the battery would be recharged routinely, probably every night when electricity costs are lower. Several auto makers have announced plans to have tens of thousands of plug-in hybrids on the market in the next 2–3 years, but it remains to be seen whether consumers will be willing to pay the extra initial costs for these vehicles. Toyota claims that its plug-in Prius using a lithium-ion battery will travel 14.5 miles (23 km) on a single charge before it switches to a conventional hybrid performance. In such use Toyota expects to get 134 miles (215 km) per gallon of gasoline.[1]

An analysis by the National Research Council estimated that a government policy to encourage purchases of new plug-in hybrids would require large subsidies of hundreds of billions of dollars,[2] but of course the picture would change entirely if current research efforts led to low cost, light weight batteries with very high energy density. Such a result is the target of a long-term collaboration between the Advanced Research Projects Agency – Energy (ARPA-E) of the US Department of Energy (DOE) and the private sector United States Automotive Battery Consortium. The 2009 funding for this work was part of the $787 billion stimulus package appropriated by Congress.

In reviewing the hopes and doubts associated with this effort, it is worthwhile to recall what gains might be anticipated from a shift toward greater dependence on the battery and less on the gasoline engine. First, since electricity is generated in large power plants there is the advantage of a smaller number of places of action away from the many millions of automotive point sources. This centralization allows for use of fossil fuels that are relatively less productive of greenhouse gases (GHGs), and it would provide locales where the carbon dioxide (CO_2) could be removed from effluent gases for sequestration, should such a technique prove to be feasible. It would fit well into any expansion plans for nuclear energy. Furthermore, any electricity generated by renewable technologies could serve our transportation needs by being directly integrated into the electric grid; and at least as important, the political and economic benefits of reduced dependence on imported petroleum would be enormous.

6.2 Syngas and Liquid Fuels

For some applications we need not only to convert other forms of energy into fuels, but, more specifically for transportation, into liquid fuels. These are needed for automobiles and airplanes today and will be needed at least into the near foreseeable future. We have already described the production of alcohol by fermentation of sugar, corn, or other non-food crops, as well as its merits and demerits. Here we turn to an alternative approach that is based on an intermediate mixture of gases called *syngas*, a shortening of the full description as synthesis gas.

Syngas is a mixture of carbon monoxide (chemical formula: CO) and hydrogen (chemical formula: H_2) that is produced either by chemical reaction with steam or by partial oxidation of any of a wide range of organic compounds, using as raw materials either coal, natural gas, or the hydrocarbons in one of the plant products that we have come to call biomass. If, for example, the methane in natural gas (chemical formula: CH_4) is the raw material, syngas is formed by the partial oxidation reaction[3]:

$$\text{Methane} + \text{Oxygen} \rightarrow \text{Hydrogen} + \text{Carbon monoxide}$$

Combustion of some of the hydrocarbon source can provide the energy needed for this reaction, but at the cost of adding to the CO_2 being liberated.

In a follow-up chemical reaction the syngas can be converted into any of a variety of liquid fuels, depending on the catalyst, temperature, and pressure conditions chosen for the reactor vessel. The best established of these is the Fischer–Tropsch (FT) process, named for its inventers, who developed it as early as the 1920s and 1930s. Because of its costs of production, the FT process has only been practical until now in special circumstances. During World War II Germany used coal to feed the FT process when other petroleum sources were cut off. In South Africa during their apartheid period, FT based on coal and natural gas was the basis of their commercial diesel fuel production. Otherwise the process has until now not been competitive with petroleum-based manufacture. It could become attractive again if non-food crops become significant sources of energy and/or hydrogen was produced using renewable energy in a way that did not increase the CO_2 burden in the atmosphere.

A pictorial flowsheet of a factory that would use coal, natural gas, or biomass to produce liquid fuels via syngas intermediates is shown in Figure 6.1.[4] The box marked as *gasification* represents all the sub-steps that produce the syngas intermediate as well as the by-product CO_2. A further step serves to remove impurities from the stream and to separate the CO_2 for sequestration (yet to be made into a commercially viable step). The *FT reactor* with follow-up purification completes the conversion into a usable liquid fuel. Each of the steps in the overall synthesis would, of course, require detailed developments of the chemical engineering operations including gas purification before the reaction and the needed product separations after the reaction, all major contributors to the relatively high cost of this manufacturing process. The same figure also illustrates an alternative use of syngas: that of manufacturing gasoline by the Mobil process via the intermediate of methyl alcohol (methanol). The preferred route will depend on the economics at a given time and place.

In a sharp departure from the techniques that use either petroleum or biomass as starting materials, the DOE has offered support for a new research program to explore an alternative approach to making liquid transportation fuels.[5] The objective is to use "microorganisms to harness chemical or electric

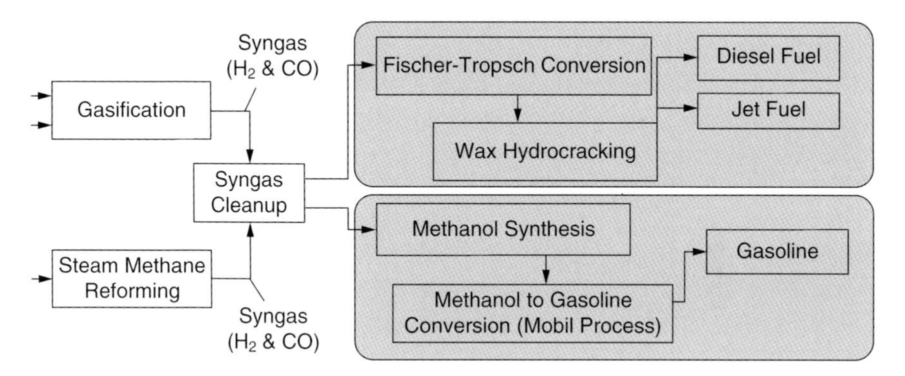

Figure 6.1 Manufacture of liquid fuels via syngas.

energy to convert carbon dioxide into liquid fuels." Success in this endeavor may be quite distant, but the benefits would be very large, including among others the ability to recycle carbon dioxide produced from conventional fuel combustion.

6.3 Hydrogen Gas

For several decades some have entertained the dream of a "hydrogen economy," of storing energy as an elemental combustible gas and using the gas for every need from industry to transportation. Since the combustion of hydrogen produces water as its only product, it appears at first glance to be a non-polluting answer to the problems of both fuel supply and global warming. But since hydrogen is only found in chemical compounds and not in its free gaseous state, any careful evaluation of these prospects must examine how hydrogen gas is to be manufactured. The current method of production is based on the reaction of steam with hydrocarbon sources from either petroleum or natural gas. If one uses natural gas, for example, the reaction is called *steam reforming*:

$$\text{Methane} + \text{Water} \rightarrow \text{Hydrogen} + \text{Carbon monoxide}$$

To produce even more hydrogen, this reaction is followed by the so-called *water gas shift reaction*, which uses additional steam to react with the carbon monoxide from the first step:

Table 6.2 Chemical equations in hydrogen manufacture.

Methane + Water → Hydrogen + Carbon Monoxide
$CH_4 + H_2O \rightarrow 3\,H_2 + CO$
Carbon Monoxide + Water → Hydrogen + Carbon dioxide
$CO + H_2O \rightarrow H_2 + CO_2$

$$\text{Carbon monoxide} + \text{Water} \rightarrow \text{Hydrogen} + \text{Carbon dioxide}$$

This very efficient industrial process produces many millions of tons of hydrogen gas each year, but it is important to emphasize that it also produces CO_2 to the same extent that it would if the natural gas were burned directly as a fuel. For reference the chemical formulas for steam reforming and for the water gas shift reaction are summarized in Table 6.2.

In view of this CO_2 production it is clear that as long as hydrogen is manufactured from hydrocarbon raw materials, its use is not a remedy for global warming. Furthermore, since we already have in place an excellent system for transporting natural gas, there is no benefit to be expected from an economy based on hydrogen unless there is found an efficient way to make hydrogen gas that does not depend on hydrocarbons.

If we are not to use hydrocarbon sources for obtaining hydrogen the obvious target source is water, which can be split into its elements (hydrogen and oxygen) by electrolysis:

$$\text{Water (liquid)} \rightarrow \text{Hydrogen (gas)} + \text{Oxygen (gas)}$$

The needed electricity could be produced by a nuclear power plant or from any renewable source, whether tides, direct sunlight, or its derivatives in wind, waves, or non-food crops. This scenario raises two questions: first, how efficient is the process of electrolysis; that is, what fraction of the available electric energy actually gets stored in the hydrogen that is made? And second, is there a better medium for storage of the electric energy than in hydrogen gas?

As to the question of efficiency, it is generally recognized that with current technology water electrolysis does not convert enough of the electric energy into the chemical energy of hydrogen to be economic. The wasted electric energy is lost as heat. Conversion figures in the range of 50–70% are commonly quoted,[6] where the energy lost in generating the electricity is not included. The results depend sharply on the catalyst used for the anode and cathode in the electrolysis cell. Current practice is to use platinum alloys for this, but there is hope for improvements if better catalyst surfaces can be found.

One of the attractive features of hydrogen as a fuel is the relative ease with which it can be oxidized in a *fuel cell*, a device that converts chemical to electric energy, in effect reversing the change accomplished by electrolysis. Fuel cells can be small enough to be used in automobiles or in individual homes, although storage of hydrogen gas may be a limiting factor. For automotive applications, conversions as high as 40–60% have been claimed in transforming the energy in hydrogen into electricity,[7] but specifics depend on conditions of operation and the type of fuel cell used. Again, the materials of construction are critical and this leaves room for optimism about future trends, but how far away is the "future"? A realistic assessment of the prospects for a hydrogen fueled car must also include the downsides. An evaluation by the US DOE[8] cited as negatives the cost of vehicle fuel cells, the inability to store large volumes of hydrogen, the absence of a carbon-free way to generate hydrogen, and the lack of a nation-wide refueling infrastructure. The DOE evaluation concluded that the hydrogen future is more that 20 years away and that as a consequence it deserves only a low priority among the alternatives.

6.4 Pumped Water or Compressed Air

Hydroelectric power has long been the most successful example of renewable energy. The potential energy associated with a water source at an elevated level can also be used to store energy that has come from other sources, simply by creating a depository lake by pumping from a lower to a higher level. The pumping technology is fully developed and can in principle use power from any source, renewable or otherwise, but it is most attractive for its ability to use renewable energy that is not dependent on conventional carbon-releasing sources.

The same idea of storing energy in potential form can also be applied in the form of pressured gas storage. Energy available at any given time, whatever its source, can be used to compress air. When the pressure is released through a turbine or other machine, the air flow converts the potential energy into kinetic energy and then into the desired electricity or mechanical motion. For sizable energy storage the air pressure chamber must be an underground geological formation such as a depleted salt mine or depleted natural gas reservoir, competing for space with the proposed sequestration of CO_2. On the positive side, however, underground compressed air has the advantage over water depositories of not affecting the local population or ecosystem above ground. A further advantage is the reduction of fossil fuel use, coming about as follows. When the compressed air is released it cools as it expands and must be warmed before it enters a turbine. This is most easily accomplished by mixing the air with natural gas and igniting the mixture; the hot combustion products are then used to drive the turbine in a conventional manner, but saving about two-thirds of the fossil fuel that would otherwise be used for the same electricity generation.

In considering new installations it is useful to make comparisons with familiar existing choices. Pressure vessels generally have longer useful lives than batteries and are made of non-toxic materials. They are, however, more expensive to make for each unit of energy storage. For use where a constant intensity of supply is an advantage, batteries are preferred over compressed air, because the battery operates at constant voltage whereas a compressed air reservoir losses pressure as it is discharged. On the other hand, power from compressed air can be delivered more quickly, a decided advantage where rapid acceleration is a plus.

To date there exists only one large compressed air energy storage system, located since 1991 in McIntosh, Alabama with a rated capacity of 110 MW. Support from the DOE is expected late in 2009 for proposed compressed air storage units that will more than triple the existing capacity.

6.5 Hot Water or Molten Salt

Excess heat is all around us. The sun heats our rooftops and all exposed surfaces, machines of all kinds dissipate heat from friction, and refrigerators and heat pumps move thermal energy from inside to out. Rooftop solar panels that harbor pipes can collect the incident radiation of sunlight and convert the energy into thermal storage in the form of hot water. Such installations are

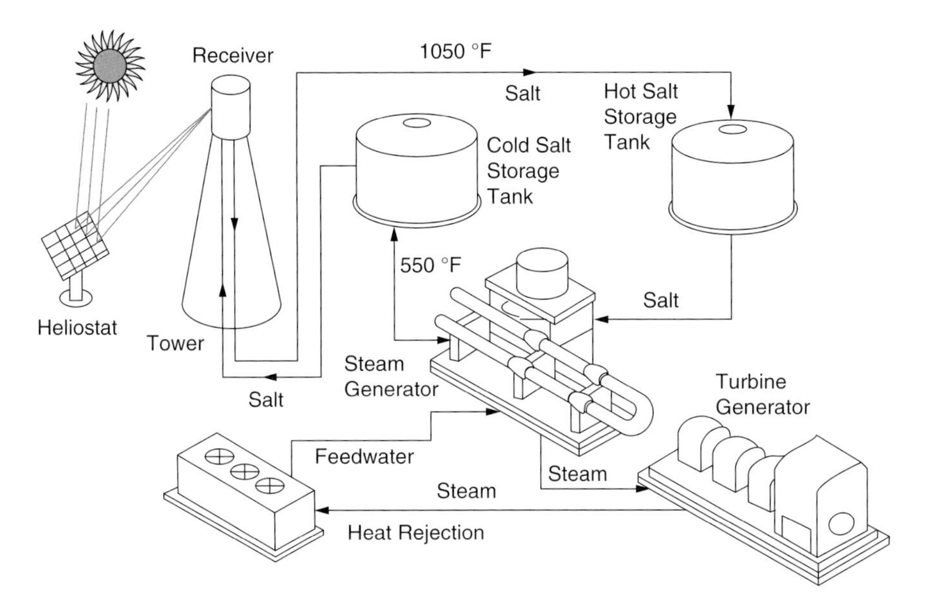

Figure 6.2 Flowsheet showing how solar energy could produce electricity via intermediate thermal storage in molten salt.
Source: Concentrating Solar Power Program at Sandia National Laboratories

commonly used for domestic hot water and/or space heating in homes as well as stores and factories. If the water can be heated to a high enough temperature to form steam, the system could be used to drive a turbine and generate electricity on a larger scale, but such a combination would require a high pressure collector and an array of mirrors to focus the radiation from a broad region on the smaller collector surface. A variation on this theme uses molten salt instead of water as the working fluid, in order to make feasible a higher working temperature that leads to more efficient conversion to electricity. A pictorial representation of such a molten salt system is reproduced in Figure 6.2.[9] The essential elements of the complex are the large field of focusing mirrors, the central heat collector, the circulating molten salt lines and storage tanks, and the final energy converter for using the heat to drive a turbine–generator combination. As long as the holding tanks for the heated carrier fluid are well insulated, the stored energy can be held for hours or days to be available when demand is high, even during the times when the solar and wind sources are down.

6.6 Flywheels

Although it has not yet found wide application, mention should be made of the possibility of storing energy in the form of energy of motion in a spinning wheel. Whatever energy is available (usually electricity) can be used to increase the velocity of the wheel and is thereby stored as kinetic energy. When the energy is to be removed from this storage device, it can be linked to an electric generator or to any mechanical drive, as desired. An interesting example of such linkage has existed in Yverdon, Switzerland for many decades in a short-run bus line. The flywheel is carried on the bus and engaged with the wheels when movement is called for, and the wheel is brought up to full speed at each depot. The system works well because the bus route is short and relatively little energy storage is needed for each trip. In this connection it is important to note that modern carbon-composite filament technology has made it possible to form light weight spokes of greatly augmented tensile strength, thus making possible higher speeds of revolution and correspondingly greater density of energy storage. Since the energy stored increases as the square of the rotation speed, such changes should make flywheel forms of storage more efficient and therefore more attractive economically. Experimental units now include magnetic bearings and vacuum chambers to reduce friction losses to a minimum, and the recovery of stored energy has been reported to be as high as 90%.[10] Rated power for existing units are still quite small, in the range of 45 kW.

In comparison with other energy storage devices, the essential advantages of flywheels are to be found in their fast response (minutes), long lifetimes (decades) with multiple cycles in and out, and low maintenance. With respect to batteries, in particular, the flywheel technology could overcome the shortcomings of low capacity and long charging times.

Notes and References

1. Hiroko Tabuchi, "As Rivals Gain, Toyota Announces Plug-In Hybrid Prius," *New York Times,* December 15, 2009, p. B5.
2. Jad Mouawad and Kate Galbraith, "Study Says Big Impact of the Plug-In Hybrid will be Decades Away," *New York Times,* December 15, 2009, p. B5.
3. This translates into the chemical formula:
   $2\,CH_4 + O_2 \rightarrow 4\,H_2 + 2\,CO$

4. Wikipedia article on synthetic fuel, http://en.wikipedia.org/wiki/Synthetic_fuel.
5. Jeffrey Mervis, "ARPA-E Puts Another $100 Million on the Table," *Science*, December 7, 2009, http://bit.ly/7xaug5.
6. Linus Pauling, *General Chemistry*, Section 15-2 (San Francisco: Dover Publications, 1970).
7. Fact sheet published by the Department of Energy (DOE) Energy Efficiency and Renewable Energy Information Center, "Hydrogen Fuel Cells," http://www.hydrogen.energy.gov/pdfs/doe_h2_fuelcell_factsheet.pdf.
8. Robert F. Service, "Hydrogen Cars: Fad or the Future," *Science*, Vol. 324, June 5, 2009, p. 1257.
9. Figure 6.2 is from "Concentrating Solar Power Program at Sandia National Laboratories," http://www.sandia.gov/Renewable_Energy/solarthermal/NSTTF/salt.htm.
10. Alan Ruddell, *Storage Technology Report: WP-ST6 Flywheel* (Didcot, UK: CCLRC-Rutherford Appleton Laboratory, June 17, 2003).

7

The Negotiating Process

We have already pointed to the necessity that the states within the international system reach cooperative agreements that resolve or at the very least significantly diminish the linked problems created by a carbon-based energy system and continued global warming. The optimism generated by the belief that the market and human ingenuity would respond quickly and effectively to these challenges is not in full retreat, but even its most forceful advocates now recognize the possibility that the market will not respond quickly enough or effectively enough to energy scarcities and environmental deterioration. Yet, while some uncertainties still exist on the details about what is happening, how quickly it is happening, and what can or needs to be done, there is also growing recognition that ecological changes in some areas could be

The Challenge of Climate Change: Which Way Now? 1st edition.
By Daniel D. Perlmutter and Robert L. Rothstein.

irreversible, and that we know enough to see the need for cooperative efforts to create a new energy economy that can impede or diminish the growth rate of global warming.

Unfortunately, knowing what we *ought* to do and what kinds of insurance we *ought* to be buying against various orders of catastrophe are hardly equivalent to doing so. We have already discussed some of the obstacles to cooperation at the domestic and international levels in Chapter 2. That they remain formidable is obvious and may account for the growing emphasis on various scientific–technological–engineering alternatives that might mitigate warming effects, if cooperation continues to flounder and the energy and environmental crises continue to intensify.[1]

7.1 A Period of Transition

It has become a commonplace to assert that we are in a period of transition between one energy economy and another entirely different, if as yet unclear, energy economy. Not so long ago, with the end of the Cold War, the defeat of Saddam Hussein in the first Gulf War, the acceleration of globalization, and the seeming unilateral American dominance of the post-Cold War order, this transition appeared likely to occur gradually, peacefully, and cooperatively. We were told that the triumph of democracy and capitalism signaled "the end of history," that globalization would spread prosperity beyond the rather narrow confines of the developed world, and that the conversion to democracy of increasing numbers of authoritarian states would bring not only peace but also a greater willingness to cooperate to resolve international problems.[2]

The forecasts of a benign transition to a new international system were premature: globalization brought prosperity but also rising inequality; elections in newly democratized states did not immediately provide anticipated benefits; and many authoritarian regimes resisted the desired transformations quite effectively. Moreover, as some ancient conflicts continued to fester and new ones emerged, the notion of an easy and gradual transition to a new world order disappeared. The meltdown of the world's financial architecture in 2008 and after, as well as uncertainty and anxiety about what might replace it, has obviously intensified the difficulties of transforming the world's energy economy, and of avoiding an excessive focus on short-run calculations of self-interest.

7.2 Our Worst Fears

Despite the almost self-evident need for multilateral cooperative agreements and an appropriate domestic policy response, the "scientific consensus on prospects for global warming has become much more pessimistic over the last few years. Indeed, the latest projections from some reputable climate scientists border on the apocalyptic."[3] Doubts about the feasibility of needed changes in our political and economic lives are fortifying fears about the future. Thus John Beddington, the British government's Chief Scientific Adviser, recently gave a "bloodcurdling speech" about the horrors awaiting us: "By 2030, he said, the world will be facing a perfect storm of food, energy and water shortages caused by population growth and exacerbated by climate change."[4] And James Lovelock, another scientist, predicts that "global warming will have wiped out 80% of humankind by the end of the century."[5] These are sobering assessments, even if one allows that the most pessimistic forecasts are a complex mixture of science, ideology, and personal predisposition. One hardly needs to add that the current economic collapse, whose effects may last for at least a decade, will make cooperative efforts to respond to these trends even more difficult, as will of course the continued conflict with and within some parts of the Islamic world, which is also a part of the world that controls a significant share of the world's petroleum reserves.

There are also other trends and developments that are likely to threaten an easy return to the liberalized progress of the past. The most noteworthy may be:

- The rise of China and India, strong competitors for resources, rising economic and military powers, and not always willing to cooperate in the pursuit of international public goods or to limit their use of carbon-based fuels.
- The aggressive nationalism of Putin's Russia.
- The difficulties that the European Union has had in sustaining its integrative momentum, especially with the addition of a large number of needy states with a different political culture.
- The rise of anti-Americanism under the Bush–Cheney administration, which even a more popular Obama administration will have difficulty turning around quickly.

These developments create multidimensional complexities and uncertainties about how to negotiate cooperative agreements in the midst of so many divergent interests, ideological conflicts, festering resentments, and varied levels of development. Or, to put it differently, the post-Cold War visions of benign globalization and democratization have been superseded by a potentially malign vision of renewed conflict and competition and regressive and dangerous geopolitics. In these circumstances, the familiar gap between the need for high levels of effective international governance and cooperation and the ability or willingness of either states or international institutions to provide it is likely to be wider than ever.

A similar conclusion can be derived from a narrower focus on the geopolitics of energy. For more than three decades, almost irrespective of the fluctuating price of oil, various scholars, oil experts, and political figures have warned, sometimes stridently, about the dangers that await the developed world – and, even more so, the non-oil developing countries – if major efforts are not undertaken to radically reduce dependence on an energy source that is the major factor in causing global warming, that is likely to become scarcer and more expensive, and that is largely under the control of potentially hostile and/or unstable powers. The alarms have largely been ignored because public opinion has not been roused sufficiently to change behaviors as sharp price fluctuations undermine support for the development of costlier alternative fuels. Further resistance has come from large energy companies able to pressure weak and divided governments unwilling to risk policies that might be expensive.

Repeated warnings from reputable sources seem to have no effect, except on the relatively small group that is aware of the dangers of inaction. For example, the International Energy Agency has projected that world energy output must increase by more than 50% in the next 25 years to meet anticipated demand, but that there is no certainty that such an increase will be available (especially if there is further turmoil in any of the key suppliers from war or internal conflict). The political and economic consequences of failure to meet this demand could be catastrophic.[6] With an increasing dependence on coal and a severe decline in environmental quality one could see trade protectionism ever more prevalent, sharp competition to control supplies of raw materials, violent conflicts escalating, and desperate mass migrations creating humanitarian disasters.[7] The United Nations (UN) has already produced a study of the existing and potential difficulties created by "environmental refugees," who do not fall within standard definitions of refugees and might become almost unmanageable in terms of numbers. A 1 m (39 inches) rise in sea levels along the major river systems in Asia, for example, could displace 24 million people – "paperless paupers," as they have been described. At worst, a new Cold War could emerge but with different alignments and different issues at stake, not focused on ideological superiority but rather about resource conflicts, financial instability, and a deteriorating trading system.

The difficulties of understanding where we are heading or how to get there have been compounded by one major tendency: if all the trends in the post-Cold War era at least initially seemed to point in the same positive direction, the opposite now seems true. Globalization has obviously slowed as world trade has shrunk, protectionism seems on the rise, and some degree of what could be called "deglobalization" – a retreat from increasing degrees of integration – seems to be occurring. The retreat is not total, however, and the pace of globalization may once again pick up if trade financing is available, demand is sufficiently stimulated, and the World Trade Organization (WTO) helps to resist one or another form of protectionism. Still, for the moment the states of the world have not yet definitely arrested the free fall in their economic prospects and high levels of anxiety about the future are everywhere. In short, there is an obvious danger of a fairly prolonged return to economic nationalism and "beggar thy neighbor" policy responses. Nicholas Stern has argued that policies to deal with climate change and the economic crisis could be complementary, and investment in low carbon modifications could accelerate and support efforts to stimulate growth, but political and psychological support for such actions is still limited.[8]

It is worth emphasizing that anticipating the worst and acting as if it is the most likely outcome of present trends may create policy immobility, because all choices seem so bad. The fears of the pessimists are more likely to be realized if we do nothing or retreat into a desperate each-man-for-himself posture, but it is important to emphasize that environmental determinism is simplistic, that we can still make choices that can significantly influence our fate, and that even in the worst of circumstances there are still choices that can or must be made that can ameliorate or diminish costs and dangers. We shall look more carefully in the rest of this chapter at whether there are ways and means of improving the prospects for successful cooperative agreements.

7.3 Guidance from a Theory of Bargaining

Studies of the theory and practice of negotiation have produced a large body of ideas that seek to provide guidance for successful negotiations. We shall return to this issue in more detail in the next chapter but it may be useful to provide a list of some of these propositions at this point, as a kind of analytical frame of reference. The list would include the following:

1. Any agreement must reflect the interests of the parties and all parties must feel they gain more by signing than by opting out.
2. The desire for reciprocity in future negotiations and the desire to maintain a reputation for fulfilling commitments are important reasons for the survival of an agreement.
3. The tactics of side payments, compensation, and issue linkages may induce the reluctant to sign and implement agreements.
4. Leadership by the most powerful state or states is imperative, and it may be especially important if the leader is willing to pay the collective goods costs of agreement (i.e. will initially pay more than a pro rata share of the costs).
5. The agreement must be perceived as fair and burden-sharing must not seem inequitable.
6. Credible threats may sometimes be necessary to insure compliance and avoid free riding.

There is also obviously a great fear, especially in negotiations with many participants or with ancient enemies, that the other will not comply with the terms of an agreement. It helps in such circumstances if the overlap of interests is very clear and if expectations about what the agreement will do converge. This also implies that trust in the other participants in an agreement will be higher if there is some degree of shared culture, shared experiences, and a shared history of honoring prior agreements. These characteristics are obviously in short supply in an environmental negotiating arena that groups together a wide variety of states.

Analysis is one thing, political practice may be entirely different. Thus even the best of these conditions may be present, but fail to produce the desired outcomes, if the conflicts and uncertainties noted earlier become overwhelming. In sum, if we are going to negotiate our way out of the current dilemmas, we have to look beyond the rational or rationalistic proposals of bargaining theory to build on these insights to broaden our perspective. Part of the problem is that energy and the environment cannot be isolated from whatever else is transpiring in the domestic and international systems: for better or worse, these issues are linked to virtually everything else, not least values, ideologies, the distribution of costs and benefits, political and economic power, development prospects, commercial greed, and perhaps even personal reputations.

There are two simple points to be gleaned from this discussion. The first is obvious: we must begin by recognizing *all* the obstacles to cooperation, the political and economic as well as the scientific; ignoring them or talking about the need for a mystical "political will" to resolve them could be next to useless. The second point is to recognize that the problems of energy and the environment will not be resolved by merely adding pleas from the eminent or referring to detailed scientific studies. The central issue is how to construct coalitions, both domestically and internationally, that are strong enough and committed enough to reach substantive agreements *despite* the contextual difficulties. There are risks in acting but there may be even greater risks in not acting. Better or worse choices can always be made, even in the worst of circumstances. Apologies to the doomsters, but disaster is not yet foreordained.

7.4 Useful Lessons from the Past

Ideally, what we are seeking through some combination of unilateral domestic policies and cooperative international agreements is obvious: policies that are scientifically sound, economically rational, and politically pragmatic.

Most studies of past environmental negotiations focus on the sharp differences between the success of the Montreal Protocol on ozone depletion and the dismal failure of the Kyoto Protocol on the reduction of greenhouse gas emissions. The contrasts have been discussed in many studies: in brief, Montreal was easier and more successful because the crisis was imminent, there was scientific consensus, the costs were not high (because adequate substitutes for chlorofluorocarbons (CFCs) were available), there was agreement on a simple measure to end all use of CFCs forthwith, aid was provided to the developing countries, and there were potentially severe sanctions available for violations of the agreement. Moreover, it is important to recall that Montreal had procedures for the periodic revision and re-evaluation of its terms, a key safeguard in a context of uncertainty and risk.[9] Tolba and Rummel-Bulska also note that the agreement rested on the "precautionary principle" of acting before irreversible damage to the ozone layer occurred. This is a favorite principle of environmentalists but is less favored by economists and political scientists who are not enthusiastic about taking vast and costly risks before there is more certainty that they are necessary.

In short, we had something akin to the negotiator's dream: an agreement that seemed fair to all, that was an efficient means to a shared goal, and that left no participant worse off than without an agreement. Kyoto, conversely, failed because it was dealing with a much more complex issue, there were still some scientific disagreements about causes and effects, the costs of implementation might be vast, targets and timetables were set that seemed inadequate to many environmentalists, and – perhaps above all – critical states like the United States, China, and India refused to sign the agreement. Thus it floundered and many states failed to reach even the minimal targets set.

The contrast between Montreal and Kyoto is interesting and potentially useful but it may also be somewhat misleading. The conditions that made Montreal possible – a crisis, a simple goal, and a consensual and not too costly solution – may not reappear any time soon: in effect, Montreal may unfortunately have been a one off. But the conditions that created the failure with Kyoto are all too likely to be replicated. In fact, many of the strategies and tactics that were employed during the original Kyoto negotiations are still a source of friction and debate during efforts to renegotiate its terms. We shall in this section focus primarily on some of the general issues that have arisen not only in Montreal and Kyoto but also in various other international negotiations. We shall in the next chapter seek to move from the general to the more practical, and subsequent chapters will focus on specific policy choices.

One issue that has surfaced repeatedly concerns who gets to sit at the negotiating table. There are pluses and minuses with each venue, whether the negotiations go on in large global conferences or regional conferences or self-selected groups like the Group of 20 or the Group of 8 or the Group of 8 plus 5 or "coalitions of the willing." Conditions change, opportunities emerge and evolve, and what seems optimal in one configuration may be sub-optimal in another.

In the abstract, global conferences seem the best because of high levels of participation and the implicit legitimacy of such universal or near-universal groups. They are also potentially useful for consciousness raising and for putting an issue on the international agenda, although this can be a long way from having a significant effect on practice. All the other negotiating levels can seem second-best – fewer participants, less systemic legitimacy – but they can also be more effective in generating higher levels of performance and more widespread willingness and ability to implement policies. Conversely, the grand conferences may produce only a watered-down compromise that papers over disagreements on fundamental issues and is quickly forgotten when the delegates return home. In addition, given the obvious difficulties of complex negotiations between large numbers of delegates, much may depend on whether the final document was negotiated beforehand by a small group of government bureaucrats from key countries.

In general, the developing countries are strong advocates of large conferences that provide them with a seat at the table and a greater chance of extracting more foreign aid in exchange for support. This issue of venue has not been resolved, because states with different interests may have different preferences about the best venue for negotiations. One result may be that negotiations seem to go on at all levels simultaneously, which is at the very least costly, confusing, and potentially contradictory as different standards emerge and different commitments are accepted. It may also be true, however, that since successful policymaking must go on at all levels – local, national, regional, international – a multidimensional negotiating process may be imperative, whatever the loss in efficiency. The aim, of course, is a negotiating process that is not only effective and equitable but also cumulatively reinforcing.

Another issue concerns the fact that the international system lacks institutions with the sovereign power to compel implementation of decisions and thus most agreements – especially with large numbers of participants – must be self-enforcing. One might expect a stronger bias toward implementation and compliance in agreements among states that share cultural traditions, historical experiences, and similar levels of development. There may also in this context be more concern with maintaining a strong reputation for complying with agreements. In effect, states must see a strong self-interest in fulfilling commitments and must also believe others will share that belief. But such agreements are difficult because doubts about the likelihood that others will fulfill commitments will be high (especially if the historical record is poor) and the desire to free ride will be pervasive, at least until there is evidence of widespread compliance or a credible willingness to impose severe sanctions for non-compliance. We shall return to this issue in the next chapter.

It should also be recognized that the negotiation of an agreement, at whatever level, with however many participants, and with whatever intentions of compliance, can have the effect of tilting the participants toward implementation, if only to establish a reputation for reliable compliance with terms or to reinforce the idea that violating the terms of an agreement can set a dangerously bad precedent. The negotiations at the time of the Cold War showed that even bitter enemies can see the benefits of compliance with commitments when they anticipate gains and have concern for a reputation for reliability. Syrian and Egyptian compliance with various agreements with Israel is also illustrative. A sharp contrast can be found in the Palestinian and Israeli reciprocal failures to implement the terms of the Oslo Accords.

7.5 What Should a Treaty Accomplish?

What kind of treaty will seem to be in the interest of all to ratify and implement? Short of an overwhelming crisis that takes precedence over narrow calculations of self-interest, that agreement must satisfy widely different standards of fairness. Given wide disparities in power, this means that one standard of fairness or justice may prevail, but agreements are more likely to be effective if the stronger is aware of the need to avoid the appearance (or reality) of imposing its views. This is obviously important in the present context because disparities in power, wealth, and scientific capabilities are vast, and the developing countries have reasonable grievances about who should pay the costs of reducing existing levels of greenhouse gases and devising and funding a less carbon-intensive approach to development and poverty reduction. A strong and phased contribution from most of the developing countries will become increasingly important; helping is a necessity, not a luxury.

There is also growing expert agreement that a focus on binding, mandatory provisions and explicit targets and timetables is probably unworkable, if not counterproductive.[10] Such provisions are especially problematic for developing countries that lack efficient administrative and technical capacities. In addition, given the uncertain knowledge base, the fact that standards once set may be difficult to reform and that an explicit early agreement may obstruct the later negotiation of a better agreement suggests that binding targets and timetables may, at a minimum, be premature. There is also a difficult question about how to define such targets and who should be required to do what in what period. If an agreement sets out symmetrical targets and timetables for all parties this may simplify the negotiating task and make free riding more difficult, but it is also unfair and impractical when the parties are not in fact symmetrical. Symmetry may also promote a degree of cynicism and hypocrisy: developing countries may sign for the ancillary benefits available (foreign aid, technical assistance, praise for an ostensible commitment to environmental protection) but have neither the intention nor the ability to comply with mandatory provisions. There is also the problem here that establishing "common but differentiated" responsibilities for developing countries, a standard approach in many treaties, may also become a euphemism for doing nothing or doing as little as possible.

There is an alternative view, ably represented by Nicholas Stern, that we need to act quickly and effectively by 2020 to control carbon emissions and that targets and timetables – certainly for the developed countries and other major polluters – are a critical component in this effort.[11] Developing countries would not be asked to sacrifice development goals or to accept immediate targets and timetables but would be required to do so after 2020. The latter is necessary because, if the developed countries succeed in limiting their emissions, the developing countries (including China and India) would by 2020 contribute about 70% of global emissions. Given the latter point, there may not actually be much disparity between these views about targets and timetables.

One difference is that the critics of targets and timetables have focused on past behavior, whereas Stern is focusing on current needs. This seems largely to reflect a judgment about how imperative it is to act quickly so that significant reductions are well underway before (and obviously after) 2020 – if one believes that we *must* make substantial reductions by 2020, else global warming will become an irreversible disaster, then serious targets and timetables must be mandatory and violations must be severely sanctioned. It does seem that the risks of not acting soon enough are much greater than the risks of demanding immediate efforts to reduce emissions; as such, whatever the potential problems of targets and timetables, not setting them may reflect a judgment that the risks of a limited response are bearable. In any case, the obvious compromise of phased targets and timetables for most of the developing countries (but *not* China and India) should suffice for the next decade or so.

Decentralized and flexible systems that distribute tasks according to capacity and that are relatively easier to revise thus may be more generally effective.[12] Aldy and Stavins tell us that given the significant uncertainties that characterize climate science, economics, and technology, and the potential for learning in the future, a flexible policy infrastructure built on a sequential decisionmaking approach that incorporates new information may be preferred to more rigid policy designs.[13] This implies the need for regular evaluation of performance on a rolling basis, but who should perform the evaluations and what powers the evaluators should have are unresolved and complex questions. Raustiala and Victor also note two other important advantages of a flexible, non-binding approach to negotiations: such agreements do not have to go through the politically difficult process of domestic ratification and they may permit states to make more ambitious commitments than would be feasible under a binding agreement.[14] In sum, flexible and non-binding commitments have some clear advantages in a context of uncertainty and widely differing capacities, but there are also some clear advantages to firm targets and timetables, especially if conditions deteriorate or initial targets are not being met. What is best will presumably depend on the nature of the agreement, the degree of consensus on the severity of the problem, and the apparent time left to act effectively: a complex mixture of objective and subjective judgments.

Another theme in the field on bargaining theory is the dominance of domestic politics, or at least its equal importance with the international configuration of power and interests. Costs and interests differ not only between states but also within them. Since any formal agreement must be ratified and since even informal agreements need substantial domestic support, the need to develop a "sufficient consensus" domestically is crucial; without it, there is no agreement or an agreement likely to fail.[15] Though not the only variable in developing appropriate policies and generating a domestic consensus, political leadership is clearly an indispensable variable, provided that it is leadership that avoids ideological posturing and is willing to risk some political capital to achieve progress on these issues. Domestic bargaining in the US over climate change legislation might also be facilitated if bilateral negotiations between the US and China were to prove successful, because it would diminish fears of Chinese free riding and its quest to gain an unfair competitive edge. But the results of these negotiations both before and after Copenhagen have been disappointing.

Disappointment with the Obama administration's policies on energy and climate change has been growing rapidly in the environmental community, especially because of the President's support for nuclear power.[16] However, a good part of the problem that the President faces is a result of system overload: dealing with two major wars, an overwhelming economic crisis, and the need to reform the health care system hardly leaves much time for other issues. The political work load is compounded if the issues involve high costs, potential changes in comfortable lifestyles, and strong resistance from interest groups that either fear the potential losses implicit in a new energy economy or are ideologically opposed to the very notion that global warming is a serious risk. The vicious polarization that seems to have infected political debate in the United States makes the negotiation of pragmatic compromises even more difficult. Finally, divergent political schedules in one or another key country serve to constrict political space and add yet another layer of complexity to the negotiating process.

7.6 Where We are Heading

One final point is worth making about the potential lessons that can be extrapolated from the analysis of environmental negotiations. The Harvard economist Richard Schmalensee noted that when the time span of a problem could cover centuries "the creation of durable institutions and frameworks seems both logically prior to and more important than choice of a particular policy program that will almost surely be viewed as too strong or too weak within a decade."[17] This is very sensible and we shall return to this issue in the next chapter. However, it should also be noted that Professor Schmalensee's comment may beg two important questions: first, given the limitations of our knowledge, how can we know what institutions and frameworks we should create? And second, what choices are we supposed to make within the intervening decade, especially given that we do not have the option of doing nothing until we know enough to do the (presumably) right thing?

How useful are these generalizations? They are the best that we have been able to find and they are, by and large, realistic judgments about the limits of the possible in the current political, economic and social context. Thus the advice to be flexible, to focus on non-binding, decentralized systems, to employ rigid targets and timetables only when reasonably certain that they are necessary, and to focus on establishing institutions and policy frameworks rather than specific policies that may quickly become dysfunctional are all reasonable suggestions about the proper course of action.

It is a fair question, however, whether modest goals and flexible means will suffice. For the most part, the major difference between pragmatists and alarmists is time, that is, disagreements are over when the energy and environmental problems might become much worse, not over whether they will. Given the uncertainties about how much time we have, the key issue requires an essentially subjective risk assessment: since a modest approach might not produce significant results quickly enough but a radical approach could be very costly, it could be misguided (i.e., the wrong radical approach), and it could prove to be unnecessary, which risks do we want to run and who is to make the decision in what kind of negotiating arena? If the best agreement and one that is minimally necessary are both unattainable, how can we decrease the risks of international policy failures and increase the likelihood of gradually approximating the "good enough" – an agreement that generates steady progress toward shared goals of creating a new energy economy, diminishing global warming, and acting with fairness toward the rich and the poor, and present and future generations? We shall explore these issues in the next chapter, looking more closely at the problem of compliance and exploring some ideas developed in trade negotiations and in the development of the European Union. We will examine graduation from one status to another, "variable geometry," and deepening versus widening. Ideas from other contexts rarely fit a new problem perfectly, but they can be useful in generating new insights or perspectives. We shall also look again at the question of negotiating formats and whether including all or some "stakeholders" – however that term is defined – will improve the prospects for successful negotiations and a more legitimized negotiating process. The aim throughout will be to find some practical means of diminishing the obstacles to viable international agreements.

Notes and References

1. See Lawrence Summers, "Foreword," p. xxiv in Joseph E. Aldy and Robert N. Stavins, eds, *Architectures for Agreement: Addressing Global Climate Change in the Post-Kyoto World* (Cambridge, UK: Cambridge University Press, 2007).
2. See Francis Fukuyama, *The End of History and the Last Man* (New York: Free Press, 1992), and Samuel P. Huntington, *The Third Wave: Democratization in the Late Twentieth Century* (Norman, OK: University of Oklahoma Press, 1991).

3. Paul Krugman, "Empire of Carbon," *New York Times,* May 15, 2009, p. A23.
4. Quoted in "Renewable Energy: Greenstanding," *The Economist*, April 25, 2009, p. 86.
5. Ibid.
6. See Michael T. Klare, *Rising Powers, Shrinking Planet: the New Geopolitics of Energy* (New York: Henry Holt and Company, 2008).
7. "Migration and Climate Change," *The Economist*, June 27, 2009, pp. 79–80.
8. See Nicholas Stern, *A Blueprint for a Safer Planet: How to Manage Climate Change and Create a New Era of Progress and Prosperity* (London: Bodley Head, 2009), pp. 206–7.
9. See Scott Barrett, *Environment and Statecraft: the Strategy of Environmental Treaty-Making* (New York: Oxford University Press, 2003), Joseph E. Aldy and Robert N. Stavins, eds, *Architectures for Agreement,* op cit., pp. 11–13, and Mostafa Kamal Tolba and Iwona Rummel-Bulska, *Global Environmental Diplomacy* (Boston: MIT Press, 1998).
10. See Edward A. Parsons and Richard J. Zeckhauser, "Equal Measures or Fair Burdens: Negotiating Environmental Treaties in an Unequal World," pp. 81–2 in Henry Lee, ed., *Shaping National Responses to Climate Change: a Post-Rio Guide* (Washington, DC: Island Press, 1999), and Jonas A. Meckling and Gu Yoon Chung, *Sectoral Approaches to International Climate Policy, a Typology and Political Analysis* (Cambridge, MA: Belfer Center for Science and International Affairs, Harvard University, 2009).
11. Nicholas Stern, *A Blueprint for a Safer Planet,* op. cit., p. 144ff.
12. David G. Victor, Kal Raustiala, and Eugene B. Skolnikoff, "Introduction and Overview," pp. 17–18 in D. G. Victor, K. Raustiala, and E. B. Skolnikoff, eds, *The Implementation and Effectiveness of International Environmental Commitments: Theory and Practice* (Cambridge, MA: MIT Press, 1998).
13. Joseph E. Aldy and Robert N. Stavins, eds, *Architectures for Agreement*, op cit., p. 10.
14. Kal Raustiala and David G. Victor, "Conclusions," p. 685ff in D. G. Victor, K. Raustiala, and E. B. Skolnikoff, eds, *The Implementation and Effectiveness of International Environmental Commitments* (Cambridge, MA: MIT Press, 1998).
15. "Sufficient consensus" was a term initially used to describe the decision by both sides in the peace negotiations in South Africa in the 1990s to go ahead without unanimity on either side but with "sufficient" support – a relatively strong majority – to sustain the peace process.
16. See John M. Broder, "Environmentalists Cooling on Obama," *New York Times,* February 18, 2010, p. A16.
17. Quoted in Joseph E. Aldy and Robert N. Stavins, eds, *Architectures for Agreement*, op. cit., p. 355.

8

From Theory to Practice

A meeting of the World Business Summit on Climate Change in May of 2009 listed the following topics for discussion: "how to create robust global carbon markets, finance clean energy, boost energy efficiency, promote investment in technology, underpin technological collaboration, protect forests and develop sustainable land use, and manage and fund adaptation to climate change."[1] This is, needless to say, a rather extensive agenda for a 3-day conference. Still, one is struck by the apparent assumption that the solutions are or will be essentially technical and that the political problems of reaching agreement on these issues can be left for others to resolve. Given the profound political problems that exist domestically and internationally, this approach is not sensible; merely holding more meetings, large or small, to bring the

The Challenge of Climate Change: Which Way Now? 1st edition.
By Daniel D. Perlmutter and Robert L. Rothstein.

same experts together to make the same points they have already made, sometimes repeatedly, will not suffice. If the central questions are about "what to do, how and when to do it, and who pays for it" the political process will be at least as crucial as the scientific, the technological, and the economic.[2] A strong turn toward practical results is clearly needed, and is the focus of this chapter.

8.1 Different Regimes and Perspectives

The leadership variable is obviously crucial in determining or strongly affecting whether environmental issues rise on the political agenda, but there are other issues here that deserve comment. Some analysts have argued that democratic states will be unable to deal with the problems of energy and the environment because they are dominated by powerful self-interested lobbies, and a culture that is materialistic and focused on short-run gratifications.[3] There is no persuasive evidence, however, that authoritarian states have performed or will perform any better, or that closed societies will be more effective at generating technological innovation, or that grassroots mobilization of civil society in pursuit of shared goals will be possible in authoritarian societies that have deliberately undermined all the elements of an open society. This is not to deny, of course, that democratic regimes have some notable deficiencies, not least the tendency to inconsistency from frequent elections and changes in government. Nevertheless, most democratic states are far richer and far more effective in performing the crucial tasks that will be necessary to deal with energy and the environment, such as planning for the future, subsidizing alternatives, and nurturing scientific and technological research. It remains to be seen whether contemporary China will be an exception to this generalization. While maintaining strict political controls, China has thus far successfully shifted from a closed communist system to a semi-open mercantilist economy that is heavily investing in renewable energy – jumping the queue, as it were. We shall return to the China issue in the concluding chapter.

The ability to reach successful international agreements on energy and the environment is closely linked to the prior creation of a domestic consensus on these issues. That ability, however, obviously varies widely across the universe of states that are major contributors to global warming. Consensus can be manufactured or imposed in authoritarian states, although even here public and elite support will presumably affect how effective that consensus will be in practice. The variations among democratic states in achieving a sufficient consensus may be narrowing, in part because of a growing scientific consensus and in part because of changes in government in the US and elsewhere. Nevertheless, the differences remain substantial and are likely to remain so for some years because of fears of rising costs associated with a new energy economy, fears of losing competitive advantage if one adopts new processes and competitors do not, and a kind of generalized fear and anxiety about economic futures, both individually and nationally.[4]

Even with strong political leadership and new knowledge about the pace and depth of climate change, we are likely to see continuing tension between unstable domestic consensuses and divergent interpretations of national interests and a growing need to achieve substantively significant degrees of international cooperation. Democratic governments, whatever their abundant virtues, inevitably add a large degree of unpredictability to the negotiating process, since no one can be certain what the next administration will or will not be able or willing to do. In any case, if global warming continues, regime type may not be a very significant variable: deteriorating conditions may force *all* states to take very similar measures, many of them mandatory, to adapt to or to mitigate the most severe threats.

There is yet another issue that has emerged in the negotiating process. This concerns the effects of the different perspectives of the economist, the environmentalist, and the political scientist. Put simply, the environmentalist is prone to focus on the "precautionary principle," a form of "worst case" analysis that insists on acting before all the evidence is in for fear that delays in action will create an irreversible slide into catastrophe.[5] We are urged to do things "regardless of cost," a position that does not make sense to economists who insist that we cannot or should not ignore costs and that we should try to retain flexibility in light of the limits of our knowledge. One ought also to recognize, as Paul Krugman does, that it is necessary to ignore what he calls "junk economics": he decries simplistic arguments that "protecting the environment would be all gain, no pain," but also argues that "the best available estimates suggest that the costs of an emissions-limitation program would be modest, as long as it's implemented gradually."[6] More detailed arguments about the costs of developing and implementing a low carbon economic strategy have been presented by Stern, who estimates the costs in the range of 1–2% of GDP (gross domestic product) in the decades ahead, surely less than the costs of dealing with intensified global warming. In any case, the conflict between the "either/or," yes-or-no focus of the environmentalist and the "more or less," let's weigh the costs and benefits focus of the economist (and the political scientist) can be quite stark and can lead to sharp differences about what should be done and when.

Finally, the political scientist is likely to resist any approach that is essentially apolitical, that focuses too narrowly on finding the appropriate institutional mechanisms to overcome collective action problems, and that ignores the feasibility of the political settlement that the institutional mechanism is meant to defend.[7] From this perspective, compromises and trade-offs are inevitable and ignoring them may create another case of the "best being the enemy of the good" – especially when we are not even very sure about what the "best" is. In any case, all of these professional biases may make cooperative agreements very difficult.

8.2 Improving the Prospects

Whether local, regional, or international, the effects of global warming are expected to vary in magnitude as well as geography. Furthermore, because the abilities to deal with either the sources of the problem or the consequences of climate change also vary greatly, the willingness or ability to pay the costs of mitigation and adaptation are rarely symmetrical. The Holy Grail of a stable global policy framework, effective international institutions, and a flexibly adjusted and mutually agreed upon price for carbon emissions is thus intrinsically difficult to achieve. There are multiple questions to answer: at what level should negotiations begin; who gets a seat at the table; will the form of an agreement make any difference; what approach to the policymaking process is most likely to produce substantive agreements; and what principles should provide guidance in answering these questions? We begin by summarizing some of the practical conditions usually associated with successful negotiations. They are frequently ignored and why this is so should concern us.

In the first place, governments must be convinced that they have a strong interest in negotiating and paying the costs of an international agreement and, of course, they must have a strong domestic coalition willing to support whatever agreement is achieved. One obvious implication is that there must be negotiations on at least three levels, sometimes simultaneously, sometimes consecutively: the domestic level; the larger group level within the international negotiating system; and at the intergroup level to reach a final agreement. Each level has problems of its own, which compound the difficulties at the final intergroup level. This complex negotiating process, especially within the United Nations (UN system), has become increasingly dysfunctional: intragroup consensus has always been fragile and the ensuing intergroup negotiations have frequently produced only masterpieces of calculated ambiguity. The process survived only because of the absence of an effective alternative for negotiations between more than 150 states.

The obstacles to progress at the national level include divergent political schedules, perhaps because of forthcoming elections or leadership changes. The international process can be excessively protracted because of internal splits within each group (especially the very numerous and very diverse Third World countries) and because there may not be much overlap between the package of demands each group produces. Disagreements about fairness may also delay agreement, as will uncertainties about the distribution of gains and losses. One seeks, of course, agreements that produce mutual benefits (or, if necessary, compensation for losers) but differences in power and perceived interests can affect calculations of acceptable terms and the proper allocation of costs. The obstacles may be more easily surmounted if a rich and powerful state is willing to pay more than a proportionate share of the initial costs of an agreement to insure a more beneficial future outcome, but such sacrifices are difficult in the current political and economic environment. These obstacles exist in all negotiating arenas but they are most severe at the level of global conferences.

If consensus exists within a policy area, the domestic arena will usually be less consequential. In any case, domestic policy, once established, is hard to change because coalitions emerge to defend it and to oppose new policies with different configurations of winners and losers. Decisionmakers are also likely to become identified with existing policies, except when new to office, to ignore contrary information, and to rationalize failure as merely a preface to future success. Moreover, in contrast to rational actor models of decisionmaking, the decisionmaker in the real world is likely to be uncertain of goals and priorities, biased in his or her choices, pulled in a variety of directions by emotion and reason, and to be heavily stressed by the need to make choices that are bound to leave some important constituencies unhappy.[8] Domestic conflicts are important in and of themselves but they may also undermine the international negotiating process.

Another problem with this condition is that measuring success is not completely straightforward. Success (or failure) reflects both objective and subjective judgments. Thus moving toward the achievement of some goals – say, meeting targets for emission reductions – may be measured objectively. But subjective judgments are also involved, since the various parties to an agreement must feel that the agreement is successful according to their own standards. Even if some goals appear to have been objectively achieved, the question of whether they are adequate (at least in terms of the present state of knowledge) is or can be largely subjective. There is the additional complication here that short-run successes or failures can become long-run failures or successes. In short, premature judgments of success or failure may be misleading or destructive.

A second factor facilitating the negotiation of international agreements is likely to be a widespread sense, especially among key participants, that the parties will comply with the agreement, that they have the ability to do so with or without credible and serious sanctions for non-compliance, and that they have not signed with the intention of cheating. There is serious debate among professionals about whether non-compliance should be treated as a violation to be sanctioned or a problem to be overcome by persuasion and an effort to build the capacity to comply. Abstract answers to this question are not, however, very useful. Much depends on the seriousness of the violation (for example, severe in the case of Iran and the Non-Proliferation Treaty), the importance of the violator in the context of the issue at hand, and the administrative and technical capacities of the party that fails to comply. The ambiguities are obvious: thus China failing to comply with an agreement to reduce carbon emissions is a very serious violation but the attempt to impose severe sanctions might be ignored or lead to China walking away from any further commitment to reduce emissions. But if sanctions are not applied or at least credibly threatened what incentive will China have to comply in the first place, apart from a growing realization that global warming will be as disastrous to China as it will to all other states? Conversely, an African state that does not comply because it lacks the capacity to do so should not be a candidate for sanctions: its failure may not have more than local consequences (which might be severe) and it needs technical and financial assistance to acquire the ability to comply. In both cases, short of an emergency that arises from new knowledge about global warming, persuasion and assistance seems the more sensible model for efforts to deal with non-compliance. Or perhaps one should conclude with that famous economic response to all questions: it all depends. And what it depends on here is judgments about the seriousness of the violation in context: for example, if global warming continues to escalate, non-compliance by a large country or even many small countries could constitute a mortal threat that justifies severe sanctions. The same sort of contextual judgment is necessary in regard to free riding: it all depends on who is doing it and how serious an issue it is at the moment. Furthermore, free riding may be held in check by awareness that non-compliance will affect relationships that will continue into the future. Compliance will also be more likely if sanctions for non-compliance are transparent and consistent and if a dispute settlement mechanism exists to resolve conflicts.

Compliance and implementation are also likely to be strongly affected by the play of domestic politics. One problem is that the individuals who negotiate an international agreement are unlikely to be the individuals responsible for implementation. The sense of "ownership" of the agreement may dissipate and a different set of interests and perspectives may come in to play. In addition, the conditions prevailing when agreement was achieved frequently change, sometimes fundamentally, and governments come and go with some regularity. Uncertainty about the willingness or ability of other parties to an agreement to implement it faithfully is thus endemic and may indeed generate a vicious cycle downward – pre-emptive defections anticipating pre-emptive defections. Agreements that are not implemented may be worse than no agreement at all in that they may make future agreements even harder to negotiate.

There is another practical reason to take the implementation issue seriously. The vast majority of negotiations will be between national governments in either a bilateral or multilateral setting. But the agreement negotiated will frequently be implemented at the sub-national level, since regional and local authorities may have control over such matters as energy production, transport networks, and enforcement of regulations. To illustrate, the UN Development Program has estimated that sub-national governments influence "50 to 80 per cent of projects to cut greenhouse gas emissions, rendering meaningless any deal at Copenhagen unless governments put in place necessary structures to ensure commitments are translated into actions at the local level."[9] Regional and local governments also frequently control crucial financial resources and are also likely to be first affected by environmental disasters and first in charge of rescue operations. In addition, their expertise and local knowledge may be superior to that of national governments.

These considerations suggest that local capacities and local involvement ought to be a critical concern in the international negotiating process. They rarely are, in part because national governments and bureaucracies are so concerned to protect their own turf. In developing countries, of course, local infrastructure and expertise are usually minimal, thus guaranteeing policy failures unless international institutions are able to help rapidly and effectively. One ought also to recognize that, while local and regional implementation is a crucial concern, it will always be difficult to integrate this concern within an international negotiating context.

Finally, there is another condition that may in some cases have an effect on the likelihood of negotiating successful agreements. This factor concerns the design of the agreement itself.[10] With uncertainty very high, with knowledge evolving very rapidly, and with interpretations of interests potentially unstable, it is not surprising that states may seek insurance against the possibility that today's commitment can become next year's albatross. One way to deal with this situation, as noted earlier, is by building flexibility into the provisions of the agreement by establishing frequent periods of evaluation and review and perhaps by creating a compliance system that favors persuasion and assistance rather than harsh sanctions. The conditions that prevail when the initial agreement is signed are virtually guaranteed to change in important ways and it makes sense to take this into account from the start: trying to force compliance with provisions that are out of date or redundant hardly makes either political or economic sense.

These generalizations provide a potentially useful warning about what negotiators need to keep in mind: they do not guarantee success but they may facilitate it. In the context of energy and the environment, however, where politics and power, science and technology, economics and ecology, and the domestic and the international mix together in complex ways, we may need more: specific and practical suggestions about how to navigate through a difficult negotiating terrain that is constantly evolving.

8.3 The Debate on Venues

What is the best setting for negotiating cooperative international agreements? How we answer will affect the kind of international agreement that emerges and the likelihood that it will be implemented. Is it at the global level, bringing together virtually all the states of the world, the relevant international organizations, a wide variety of non-governmental organizations, the media, scholars, and activists in one place for a reasonable period of time to produce a treaty that promises to resolve or diminish the problems arising in one area of concern? Or should one seek agreement in smaller venues, say, only among those with a strong and direct influence on a problem (for example, the major producers of carbon emissions), or in already constituted smaller groups like the Group of 20 or the Group of 8 or the Group of 8 plus 5 (China, India, Brazil, Mexico, South Africa) or in specially constructed "coalitions of the willing"?

No single venue is likely to suffice for all issues, trade-offs will be imperative, and any or all settings may be good for some issues and completely inadequate for others. In addition, the negotiating process must exist as a continuum: no level, from the grassroots to the global, is irrelevant and a successful policymaking process will seek to insure that actions and policies at each level converge and supplement actions and policies at other levels. If interstate negotiations fail to produce effective agreements, and if each state is more or less left to its own devices, we may need to rethink the entire international policy process. The process may need to be stood on its head: rather than focusing on multilateral agreements, we may need to emphasize the imperative for each state to develop its own legally binding targets and timetables and its own plans for rapid responses to a crisis. At a pre-Copenhagen meeting the Australian representative suggested, for example, a series of "national schedules" as an alternative to the Kyoto approach. Such national programs might then serve as the basis for a subsequent effort to devise a common global framework of consensually agreed policies.[11] The national programs could be prototypes for what is minimally negotiable multilaterally. This is a second-best strategy and it is as yet premature: the multilateral track is difficult but not yet hopeless. But it may become so, and we need to think about alternatives now.

We shall discuss what happened at Copenhagen at the beginning of the final chapter. It suffices here to note that despite the most extensive series of preparatory meetings in the history of conference diplomacy, Copenhagen failed to meet hopes or expectations. The search for alternatives is already underway but is as yet inconclusive because of clashing interests.

The benefits of any of the smaller venues for negotiation should be clear. By bringing together the major developed countries and the major "emerging" powers (China, India, Brazil, and one or two others), with some limited representation for the other developing countries, it may be easier to reach a consensus among relatively similar parties, and it is more likely that they will have the capacity to implement agreements. Of course, China, India, and

Brazil may make the achievement of consensus difficult but the same holds in larger settings. Moreover, some developing countries who are chosen to represent the developing countries as a whole or a region within the Group of 77 (the Third World's caucusing group within the UN system) may feel the need to be more radical and demanding than they might want to be on their own in order to satisfy the wide-ranging interests of a very disparate Third World. Still, these settings are relatively more manageable than global conferences and they bring together the leading players responsible for the vast majority of carbon emissions. By the same token, they lack the legitimacy of the larger settings, some of the countries left out or only represented by others may become major "carbon criminals" in the years ahead, and those not present may not feel that any agreement is fair or reflects their interests. In short, narrower agreements among the rich and powerful and a few leading developing countries may be relatively effective in the short term but increasingly ineffective in the long run. The same may hold true for the narrowest forum: bilateral US–Chinese negotiations as the two largest emitters and the two most powerful countries. If US–Chinese tensions rise, letting others into the game may be useful, especially if they are committed to seeking a mutually beneficial international agreement. These are "second-best" solutions, but they may be the best option available if "first-best" global solutions are unavailable or too costly or too time consuming to negotiate.

"Coalitions of the willing" began to be discussed in the late 1970s and early 1980s when the futility of negotiations between the Group of 77 and the developed countries became increasingly apparent. The Group of 77 theoretically united a very large and divided Third World behind common positions; in fact, the Group of 77 never really established common positions, all demands were simply lumped together in one broad and non-negotiable package, and the outcome was a stalemate that benefited no one. The quest for a more practical and effective approach to negotiations focused on coalitions of the willing that united states that shared interests in a particular area of concern, irrespective of level of development, regional location, or ideology. The major criticisms of such coalitions include the fact that they have to be reconstituted whenever a new issue arises, that new concessions may become necessary, and that habits of cooperation never develop.[12] These appear to be minor deficiencies: the same reconstitution of coalitions is necessary within large and permanent organizations, there is a major benefit of getting outside the constraints established by the rigid group system within the UN and its affiliated organizations, and it is not clear why habits of cooperation cannot develop within a coalition of the willing. In addition, given the well-known deficiencies of the UN's personnel system, which seems unlikely to be reformed any time soon and which only occasionally rewards competence over gender or geography, the government bureaucracies that would organize and implement the decisions of a coalition of the willing could be much more effective at negotiating substantive agreements.

Coalitions of the willing have several other advantages, apart from the obvious but crucial overlap of interests on a particular issue. Recall the brief discussion in Chapter 2 of club goods within the wider notion of international public goods. Club goods are non-rivalrous but – unlike conventional public goods – excludable, for example, requiring fees to enter a national park or to use a toll road. In this sense, the advantage of a coalition of the willing is that members share an interest and agree to club rules but that non-members can join whenever they are willing to accept its agreements and abide by its rules. There are no arbitrary exclusions via geography or ideology or even wealth. Above all, such smaller but not exclusive groups are more likely to be able to work out and implement common positions quickly and to adjust more rapidly to changes in knowledge or external conditions. Foreign aid and technical assistance may still be necessary for some of the poorer and weaker members of the coalition but there is less likelihood of the wholesale blackmail that frequently occurs in global arenas, where some participants will block consensus unless or until they are bought off. In any case, while such coalitions may be theoretically second best, they may also be the best available means to establish a substantive consensus among a reasonable number of states, to do so quickly, to have the capacity to implement effectively, and to allow new members to join relatively easily. This openness to new members who accept the rules of the club is an important difference from exclusive clubs such as the Group of 8 or the Group of 20.

There may also be a useful analogy here with some practical ideas that have emerged within the European Union (EU) to deal with the fact that the enlargement of the EU has added a number of poorer, less developed countries that need more time to adjust to the community's rules and regulations. One idea reflects the "widening versus deepening" debate, that is, whether the EU should rapidly expand to include all potential members in Europe or whether expansion should be subordinate to deepening the degree of integration among existing members. The EU came down rather strongly on the side of widening but in the present context perhaps deepening should take precedence. Even a coalition of the willing is likely to contain members at different levels of development and with different interpretations of national interests. Thus the coalition is more likely to survive and prosper if it seeks gradually to deepen its commitment to agreed principles but simultaneously recognizes the need for differentiated degrees of commitment and implementation: the EU calls this "variable geometry," the notion that not every member must take part from the start in every policy initiative. This recognition of "common but differentiated responsibilities" has been widely discussed among policy proposals on environmental negotiations, but it has too often become a euphemism for the evasion of responsibilities.[13] Within a coalition of the willing (or a club), however, it should be easier to establish realistic differentiations, to monitor compliance, and to "graduate" members who have developed the capacity to completely fulfill various commitments.

The idea of "graduation" emerged in international trade negotiations to indicate that when some developing countries were granted preferential access to developed country markets, they would graduate, that is lose their preferential status, when they reached a certain level of development. There has inevitably been controversy about the criteria for graduation and who should make the decision. In the present context, the idea implies that differentiated but progressively demanding benchmarks would be established to measure progress and to increase requirements as performance improves. It would also presumably be easier to harmonize domestic policies (if in a differentiated fashion) in a coalition of the willing and to coordinate national policies to avoid unilateral actions. In sum, one is seeking here a negotiating framework that has the best chance of establishing strong commitments quickly, that is likely to be more flexible but also less ambiguous than loose commitments in a global arena, and that differentiates commitments and responsibilities for states with different capacities, but not to the point of evasion.

One alternative to coalitions of the willing has been suggested in a recent book.[14] The authors advocate a Group of 16 (G-16), which would bring together the major powers and the rising powers in an effort to unite the power to act and the responsibility to do so. The fixed core of membership would presumably encourage predictability and accountability and permit a degree of institutionalized dialogue to generate consensus. Whatever the virtues of the G-16 as a policymaking center in the global arena on many issues, we do not believe that it would be superior to a coalition of the willing in regard to energy and the environment. The G-16 would lump together countries with very diverse interests and diverse interpretations of the dangers of global warming. These differences could delay, or thwart, or water down proposals for common action. Conversely, in an environment where the need to act quickly may become imperative, a coalition that unites states with the same priorities and the same sense of the need to avoid delay may be more effective.

8.4 Bargaining Strategies: Domestic and International

No international policy can succeed if it is incompatible with domestic policies and no domestic policy will work effectively unless it is compatible with international agreements. Nevertheless, state actions should and can be the first building blocks in a multidimensional policy universe, and in fact a commitment to establish voluntary national plans to reduce emissions was one of the few agreements that emerged from the Copenhagen conference. To illustrate, Hal Harvey, CEO of Climate Works, has noted a variety of actions that states themselves should undertake, actions that "pay for themselves [and] reap huge environmental dividends. And they do not require that 190 nations sign a treaty to achieve results."[15] Actions such as improved efficiency standards for appliances and buildings, or improved mileage standards for automobiles, or actions to protect national forests can be established nationally and deepened by subsequent international action.

But what is the appropriate international bargaining strategy for states? There are, of course, many bargaining models and many different approaches to conflict resolution. However, the most prominent approaches to bargaining tend to differentiate between distributive bargaining and integrative bargaining. The former is more conflictual because it focuses on who gets what share of a fixed pie, and the latter is relatively more cooperative because it focuses on increasing the size of the pie so that both sides can extract larger gains, although there can still be a conflict over who gets the bigger share of the bigger pie. As useful as these models have been in describing and analyzing some traditional issues in international relations, they do not seem useful in the present context. There is implicit in the traditional approaches an emphasis on power, on who has the capacity to earn a greater share of the pie, on what the weaker can do to improve its prospects, and on when or why the stronger might choose not to exercise its power. Competition, not cooperation, is the main focus. One should also note in this connection that some of the more elegant bargaining models favored by economists do not work well in the present context because they tend to focus on a single encounter, not on the continuity of a relationship, and because it is difficult to assign quantitative values to a range of different outcomes.

Traditional strategies remain relevant, for example if we focus narrowly on the geopolitics of energy and the conflict between oil producers and consumers, but they are less relevant and perhaps counterproductive if we focus on broader questions about cooperative efforts to create a new energy economy and to diminish the dangers of global warming. Implicitly, in these circumstances, what states seem to be bargaining about is not merely who gets what but also about establishing rules of behavior in an arena of competitive but peaceful coexistence. The aim here is to pursue and protect national interests but not to the extent that there is a threat to the international order itself or to equitable treatment of all members of a group or club. One seeks here to establish, slowly but steadily, the idea that short-run national interests should not be pursued to the extent that they threaten the stability of the international system itself. It is worth noting that China's policies at Copenhagen and its quest for competitive advantages in international trade seem to be at odds with this new bargaining perspective: its policies emphasized rather too strongly the pursuit of its national interests.

One problem here is that bargaining models that would be more appropriate for a world of multidimensional chessboards in which cooperation is essential and the games must go on for decades or more do not seem to exist. Peace negotiations in protracted conflicts suffer from the same deficiency because the aim is more than just establishing a pattern of mutual concession and gradual convergence on an acceptable compromise agreement: the weaker side is also seeking respect and recognition, both accept the need to settle conflicts in the political system, and the shared long-term aim is to establish new rules for peaceful coexistence.[16] The last point may move us somewhat closer to an improved approach to bargaining on energy and environmental issues. The framework for negotiations on these issues is unstable because there are high levels of risk and uncertainty, knowledge changes rapidly, and perceptions of interests can alter quickly. Thus a new approach to bargaining must seek to manage rapid change and to establish mutually agreed rules on how to do this and how to interpret knowledge that is continually evolving.[17] Expectations in the traditional bargaining models tend to be competitive but here we seek mutually reinforcing expectations that take both cooperation and competition as a given, which in turn should affect the knowledge and information that is sought and the styles of behavior that parties adopt toward their negotiating partners.

This movement toward an alternative approach to bargaining has barely begun. There is a possible linkage with some of the more philosophical and normative literature on the environment, which focuses on our shared responsibility to future generations to hand over a planet that has not been devastated, but the linkage is not yet clearly established. Thus a new approach is an ideal that, over time, may begin to alter perceptions and behavior. Perhaps the most that can be said at the moment is that initial efforts should probably focus on integrative patterns of bargaining: increasing the size of the pie, an emphasis on problem solving, and on convincing the rich countries that a clear commitment to equity and to development may generate cooperative responses that begin movement toward shared, practical commitments to establish new rules for dealing with energy and the environment.

8.5 Big Bang or Accelerated Incrementalism?

Should we attempt to reach agreement by a "big bang" approach – like the Copenhagen conference – or is it possible to do more with less, to rescue incrementalism from its detractors? We have already emphasized our doubts about the wisdom of relying on large global conferences to make substantive decisions but we have not explained why we feel a more modest policymaking style in a smaller forum might produce better results. Incrementalism, which assumes that the best guide to tomorrow's policy is yesterday's policy, that change is "incremental," has had a very bad press in recent years, especially in regard to failed peace processes in the Middle East and elsewhere. And much of the criticism seems justified when small steps fail to cumulate into steady progress toward a goal. Shocks and surprises can easily derail a process before much progress has been achieved, and it may be thought that genuine breakthroughs may require a grand conference (like Camp David in 1978 or Dayton in 1995) that bring all the key leaders together in order to overcome immobility and inertia. But the conditions that facilitated success in those cases are not easily replicated, especially with more than 190 countries wanting to have their interests protected.

Vertzberger has also argued that in high-risk decisions the incremental process provides a false sense of confidence about the controllability of the situation and induces cognitive rigidity that discourages learning.[18] But any decisionmaking process that involves high risk may do the same and it is not clear that a *sustained* strategy of incrementalism would not produce superior results. This is especially true if it is important to start a new policy process after earlier negotiations have deteriorated into a futile stalemate.

What the critics of the incremental approach seem to be criticizing are incremental actions that are isolated and uncoordinated, not an incremental *strategy* that begins with a shared goal that is to be gradually approached by planned steps that continue through thick and thin. Farrell and Bozon come close to what we have in mind here. They argue against "huge bets on spectacular, hard-to-attain breakthroughs" but for "a focused drive to make hundreds of incremental improvements that add up to a large cumulative impact on reducing energy consumption."[19] If they had also advocated an initial effort to establish shared goals that would provide guidance for the choice of each incremental action, and if they asserted the need to plan from the beginning to deal with inevitable shocks and surprises, their approach would have been nearly identical to the one advocated here. The point is to sustain incrementalism, to make it a long-term commitment about how to proceed, not merely a collection of easily derailed and isolated actions.

We believe that this focus would diminish most of the deficiencies normally attributed to incrementalism. There may be resistance to such an approach arising not only from the history of the previous failures of incremental policies, but also because an incremental strategy may seem insufficiently grand (or grandiose?) for the scale of problems in the energy and environmental realms. Some have suggested a "go-for-broke" gamble on one potential technological breakthrough, but grand gestures may lead to grand ceremonies on the White House lawn rather than to practical results when the time for implementation is at hand. The polar opposite of the big gamble is to focus on a comprehensive set of policy measures, each of which is adequately funded. This could be incredibly expensive unless clear priorities are established. The preference to be offered here is for gradually accelerating incrementalism, and we will lay out in the next three chapters our version of what choices to make in what time period.

We have already emphasized the dilemmas and dangers of making policy when so much is uncertain, when interests are unclear or unstable, and when the world economy is in a very fragile state. In such circumstances it may be better to be prudent, to do what can be done rather than what may or may not be necessary to do, and to seek above all to start a process that has wide support, that can be sustained, and that has the potential to do much more once a degree of policy momentum has been established.[20] One might attribute a significant (but not exclusive) part of the success of the peace process in

Northern Ireland to the sustained incremental strategy of the British government, which persisted over almost 30 years and became strong enough to withstand even brutal terrorist outrages like the Omagh bombing.

It is important not to be misunderstood here. We are not advocating an incremental strategy as a panacea or as the only policy strategy to be pursued. If the problems that we are discussing produce sharp environmental deterioration and if our choices become increasingly stark and dangerous, it may become imperative to make another effort to work out a global bargain that seeks significant results quickly. Until such a time, failure is not preordained and a sustained incremental strategy may offer the best hope of averting a dismal outcome.

8.6 Choices in the Context of Risk

The scale of the problems confronting us is monumental, they affect us all and may do so for a century or more, they raise severe moral and practical issues, and they must be negotiated domestically and internationally with knowledge that is constantly evolving and unconvincing to some. The overarching issue in such circumstances is risk, the risk in the problems themselves, the risk in the policies chosen to deal with the problems, and the potential risk to the policymaker if the policy fails.

Risk is experienced differently by different individuals in different contexts. As such, interpretations of risk are heavily affected by factors such as the problem's degree of complexity, the degree of uncertainty, time horizons, and personal attributes. The level of risk is likely to seem much higher if unanticipated events seem likely to occur, which also implies that the costs of failure could escalate. The point here in regard to energy and the environment is that we seem to face a "perfect storm" of risk: potential policy failures with severe consequences for both society and the policymaker, dangers from climate change that are becoming increasingly clear and increasingly close in time, developments that may become irreversible, and decisionmakers who will be unable to escape responsibility for failure to act in time. The standard tactics for risk-averse leaders – to decide by not deciding, to push difficult decisions into the future – should not suffice here. But as decades of crying in the wilderness by many environmental scientists attests, the gap between what the political system ought to do and what it will do is still great.

This chapter has sought to lay out some of the political obstacles that have thwarted successful international negotiations in the past and to offer suggestions about what might be done to improve the prospects for success in the future. The final test of our efforts must be practical and consonant with the existing technical and political constraints: what decisions should we make and when should we make them in light of the many obstacles that we have discussed? Our choices are presented and discussed in the next three chapters.

Notes and References

1. See "Coping with Climate Change," *International Herald Tribune*, May 22, 2009 (Special Advertising Supplement), p.1.
2. The quoted material is from "Global Challenge and Summit Theme: How to Shape a New Green Economy?," ibid, p. 1.
3. See Anthony Giddens, *The Politics of Climate Change* (Cambridge, MA: Polity Press, 2009), pp. 73–5.
4. A more recent and extensive analysis by Paul Krugman argues that any sensible solution must give all parties a self-interested reason to reduce emissions, which requires proper market incentives and direct governmental controls for problems that require immediate action. See Paul Krugman, "Green Economics: How We can Afford to Tackle Climate Change," *New York Times Magazine*, April 11, 2010, p. 39. This sort of focus on both the market and government intervention (and thus politics) may reduce some of the conflicts between economists and political scientists.
5. See Lawrence Summers, "Foreword," p. xxi in Joseph E. Aldy and Robert N. Stavins, eds, *Architectures for Agreement: Addressing Global Climate Change in the Post-Kyoto World* (Cambridge, UK: Cambridge University Press, 2007). and Alan S. Blinder, *Hard Heads, Soft Hearts: Tough-Minded Economics for a Just Society* (Cambridge, MA: Perseus Press, 1987), pp. 138–9.
6. Paul Krugman, "An Affordable Salvation," *New York Times*, May 1, 2009, p. A21; also Nicholas Stern, *A Blueprint for a Safer Planet* (London: Bodley Head, 2009), pp. 90–8.
7. See the preface to J. Samuel Barkin and George E. Shambaugh, eds, *Anarchy and the Environment* (Albany, NY: SUNY Press, 1999).
8. See Yaacov I. Vertzberger, *Risk Taking and Decisionmaking: Foreign Military Intervention Decisions* (Stanford, CA: Stanford University Press, 1998), p. 109ff.
9. Fiona Harvey, "Meetings are Global But Action is Local," *Financial Times*, September 26, 2009, p. 1.

10. See Scott Barrett, *Environment and Statecraft* (New York: Oxford University Press, 2003), and Barbara Koremenos, "Loosening the Ties that Bind: a Learning Model of Agreement Flexibility," *International Organization*, Vol. 55, No. 2, Spring, 2001, pp. 289–325.
11. See "Avoiding a Crash at Copenhagen," *The Economist*, September 26, 2009, p. 8.
12. See Stephen G. Brooks and William C. Wohlforth, "Reshaping the World Order," *Foreign Affairs*, Vol. 88, No. 2, May, 2009, p. 55.
13. Joseph E. Aldy and Robert N. Stavins, eds, *Architectures for Agreement*, op. cit., p. 359.
14. Bruce Jones, Carlos Pascual and Stephen John Stedman, *Power and Responsibility: Building International Order in an Era of Transnational Threats* (Washington, DC: Brookings Institution Press, 2009), p. 51ff.
15. Quoted in John M. Broder and James Kanter, "U.S. Climate Stand Worries Europeans," *New York Times*, November 4, 2009, p. 4.
16. See Ian E. Morley, "Intra-Organizational Bargaining," pp. 203–24 in Jean F. Hartley and Geoffrey M. Stephenson, eds, *Employment Relations: the Psychology of Influence and Control at Work* (Oxford, UK: Blackwell, 1992).
17. Ian E. Morley, Janette Webb, and Geoffrey Stephenson, "Bargaining and Arbitration in the Resolution of Conflict," pp. 117–34 in Wolfgang Stroebe, Arie W. Kruglanski, Daniel Bar-Tal, and Miles Hewstone, eds, *The Social Psychology of Intergroup Conflict* (New York: Springer-Verlag, 1988).
18. Yaacov I. Vertzberger, *Risk Taking and Decisionmaking*, op. cit., p. 33.
19. Diana Farrell and Ivo Bozon, "Demand-side Economics: the Case for a New US Energy Policy Direction," p. 60 in Kurt M. Campbell and Jonathan Price, eds, *The Global Politics of Energy* (Washington, DC: Aspen Institute, 2008).
20. Paul Krugman also favors a "big bang" approach on the principle that even a small probability of a major catastrophe should justify a massive policy initiative. We would agree in a different political and economic universe but believe such an approach is not yet politically feasible and that it is imperative to start a policy process now even if it must be initially modest.

9

Where Do We Go from Here?

Although we have not yet seen the catastrophes projected to arise from indirect effects of global warming in such matters as resource conflicts, food and water shortages, and mass migrations, the indicators of decline have worsened and the threats to our environment are closer. The changes have come at a rate exceeding the capacity of governments and international institutions to respond effectively, and the dire forecasts can no longer be safely ignored. In fact, current prophecies about the costs and consequences of failing to respond quickly enough to the dangers of climate change are much more plausible than earlier forecasts of disaster, such as appeared during the limits to growth debates of the 1970s and 1980s. As an example we may consider the recent study run by Battisti and Naylor,[1] who found that there is a greater

The Challenge of Climate Change: Which Way Now? 1st edition.
By Daniel D. Perlmutter and Robert L. Rothstein.

than 90% chance that growing season temperatures in the topics and subtropics will rise in the twenty-first century to exceed even the most extreme seasonal temperatures recorded during the prior century. Their models for the growth of major grains in these regions show yield losses in the range of 2.5–16% for each increase in seasonal temperature. These estimates may be compared to the European crop data of 2003 when summer temperatures were 3.6°C (6.5°F) above the long-term climatology: Italy's maize yields were 36% lower than the year before, France's production fell 30%, and fruit harvests were down by 25%.

There is not much new in arguments about the constraints of domestic politics and the weaknesses of the international negotiating system, about the developing countries' demands for more foreign aid, or about the inability or unwillingness of most developed country governments to take short-term risks to ward off long-run dangers. However, there is today the added obstacle of refusal among the developing countries to accept any limits on the pursuit of growth via a carbon-intensive rapid industrialization strategy.

While the time span to act has narrowed, it has not disappeared, and we may have some 10–20 years to put in place policies that can slow the rate of decline, stabilize the rise in global temperatures reasonably close to 2°C (3.6°F), and begin to establish the basis of a lower carbon energy economy. Effective steps must produce discernible results quickly, must spread the costs fairly, and must not seem excessively costly in a period of monumental budget deficits. Moreover, while there is always the hope that we may be saved from the errors of our ways by new knowledge or new technological breakthroughs, it is too risky to merely wait for such salvation while pursuing business as usual; such salvation might never come. It would be irresponsible either to give in to the prophets of doom and gloom or to give up on doing as much as we can as quickly as we can. This having been said, it is necessary to make choices on two fronts: on negotiating paths that can yield movements in the right direction, as well as on priorities among the many scientific–technological–economic remedies that have been proposed. This chapter will address aspects of all of these.

9.1 Is the Feasible Insufficient?

In setting about to get agreements on policy, it is essential to steer between two disappointing results. On the one hand we need to heed the warning of

Krugman that, while the best may be the enemy of the good, the "not-good-enough-to-work" may undermine both the best and the good (enough). We seek rather the "good-enough-to-work" that helps establish and set in motion a cumulatively successful policymaking process. On the other extreme one must also recognize that, if conditions continue to deteriorate, the aggressive pursuit of the "merely" feasible might be insufficient to accomplish any meaningful outcome.

In this context there is hope in the growing seriousness in most developed country governments about the need to devise coordinated sets of policies that begin the conversion to low carbon-intensive economies. There is no doubt that the failure of China, India, and the other developing countries to agree to limit their greenhouse gas (GHG) emissions is very disappointing, but there may be a 10–20-year period ahead of us in which their rising emissions will be damaging but not yet devastating. In addition, the "demonstration effect" of successful policy conversions by the developed countries may become increasingly important as political pressure toward cooperation increases, and the developing countries themselves may begin to realize the costs they are incurring by focusing on carbon-intensive development. In any case, what the developing countries do not do is no excuse for the developed countries to delay serious efforts to cut emissions. In this connection, the example set by the government of the United Kingdom in devising a coherent strategy to move toward a low carbon society may have wider international implications by helping to set in motion a process of change.

Because this British initiative may have far reaching consequences internationally, it is worth some attention to its details. A headline in an English newspaper recently read "Labour Orders Green Energy Revolution." Ed Miliband, the then Minister for Energy and Climate Change in Gordon Brown's government, had just set out an ambitious road map to facilitate the UK's efforts to meet its target of a 34% cut in GHG emissions by 2020, a goal that encompassed meeting 40% of its electricity needs from wind, tidal power, and nuclear power by 2020.[2] Responding to growing cross-party and public support for serious actions on climate change and energy security, the plan also aimed to put the UK on track to cut its emissions by 80% by 2050, an effort that would put the country in the forefront of developed country attempts to respond to climate change. The multitude of measures in the plan were aimed at rapidly cutting energy use in electricity generation, homes, transport, and industry by setting legally binding targets for carbon reduction and using all the measures open to the government to push and pull the process along – including carbon taxes, regulated markets, and subsidies for renewable energy research and development and cash for households that generate their own energy. In assessing this proposal *The Guardian* observed that "No other government in the world has published anything quite like this, both a collective statement of intention and a fairly detailed description of how carbon reduction might be achieved."

There are some who applaud the intention but insist that the goals are too ambitious to be achieved by 2020. Nevertheless, even if the plan is not fully realized as set down, these criticisms do not necessarily mean that the Miliband proposals are unimportant or doomed to failure. In fact, the effort by a major developed country government to enact and implement such ambitious measures may energize other governments to devise their own plans, a demonstration effect could occur if much of the plan is enacted and begins to have some beneficial effects on reducing carbon emissions, and there will presumably be some benefits from "learning by doing." In any case, merely breaking through the policy inertia of recent years may be as important as the details of any particular proposal.

There are, however, other criticisms that need to be noted, if primarily as a kind of warning for other policy planners. In the first place, as *The Guardian* has noted, this complex package of proposals is presented with "the audacious suggestion that [these reductions] can be achieved without depriving people of the comforts of their present, carbon-intensive, lives."[3] Presumably, this reflects a judgment about how much sacrifice the populace is willing to bear in a period of economic turbulence. Of course, the necessary adjustments and adaptations will throw the problem of cost into the future – and onto another administration. There is a prudential rule in operation here: don't ask too much of already burdened governments, don't expect them to take big risks or to be completely open about potential costs or unanticipated consequences. If progress becomes apparent and if the current economic crisis eases, it will become politically and economically easier to bear the costs of creating a new energy economy.

There is another set of criticisms that is especially severe, although it reverses the normal pattern of an excessive governmental focus on the short run. The current generation of politicians is faced with the consequences of three decades of failure to anticipate and prepare for diminishing supplies of hydrocarbons and increased dependence on unreliable or potentially unstable suppliers. Thus, for example, Great Britain may well run short of electricity at a bearable cost by 2015 and face the dismal prospect of power cuts, an increased use of coal, and possible blackmail by some suppliers. All of this was predictable some years ago, but little was done about it: it was too easy to ignore far-off dangers. Now, of course, action is imperative, but as *The Economist* has asked, "with gas too risky, coal too dirty, nuclear too slow, and renewables too unreliable," where will the necessary supplies come from?[4]

There are obvious strategies that the UK government will presumably adopt to increase supply and decrease vulnerability but it will be difficult to do if attention is overwhelmingly focused on future threats from global warming. The implication is that the Miliband proposals are too focused on the long run and not sufficiently focused on an imminent short-term threat. This is fair comment, but it needs also to be said that some aspects of these proposals will have beneficial effects on short-run demand for hydrocarbons and that the issue of supply security is primarily a foreign policy problem that requires a coordinated international or regional response. The problem could perhaps be eased by increasing storage capacity.

Finally, the Miliband proposals illustrate some of the acute dilemmas of making effective policy choices in a context dominated by rapidly changing scientific knowledge, severe economic constraints, inconsistent public support, and weak governments. Thus while some have complained that the proposals are too ambitious and possibly dishonest about the need for significant changes in lifestyles, others have argued that the proposals are not nearly ambitious enough. For example, a recent scientific report prepared for the Department of Energy and Climate Change has warned that severe warming (a 4°C (7.2°F) temperature rise) could occur by 2060 – not 2100, as previously forecast – with potentially catastrophic consequences for both developed and developing countries. Yet another report has raised severe doubts about the ability of the UK's liberalized energy markets to produce the necessary changes in energy policies and the consequent need for increased government intervention – which presumably implies even more domestic political and ideological conflict and a slower and more erratic policy process.[5] The point in the present context is that while supporting the Miliband proposals (and other such efforts) as important steps in a continuous policy process, one must also recognize and take account of the need to prepare to do more when necessary.

One other national response to energy management is worth noting. Denmark is the most energy efficient country in the European Union, which reflects plans established after the first oil crisis of the 1970s. The long-term effort focused on ensuring security of supply, cutting GHG emissions, and maintaining cost effectiveness.[6] The result has been a sharp cut in emissions in step with a 40% increase in Denmark's GNP (gross national product) in the same period. Denmark has also heavily invested in alternative fuels (especially wind and biomass) and, quite strikingly, has forced its industry into environmental innovation through a mixture of energy and carbon taxes, a "cap and trade" system, and strict building codes.

What inspired Danish politicians to risk the wrath of their citizens and their industries by imposing taxes and building codes? A spokeswoman for the Danish green/socialist party said: "We don't have a lot of resources. We have a welfare state that we have to keep up, so we have to think forward all the time and not get stuck in the past. That is where we get the courage."[7] Why have other countries not followed Denmark's lead or been as courageous as its politicians? Awareness of vulnerability is part of the answer, but other countries have been equally vulnerable and have barely responded, and much of what Denmark has done is not mysterious and has had many advocates elsewhere. Another part of the answer may lie in political arrangements: Denmark's parliamentary system has allowed majority governments to implement policies effectively, in contrast to the divided and polarized government of the United States. At the least, Denmark illustrates that effective national policies are still imperative and can still be reasonably effective as a hedge against international failures, such as the outcome of the Copenhagen conference.

9.2 Fiscal Measures

A number of ideas have been floated that use economic advantage or penalties, or taxes, as incentives to reduce GHG emissions. The one that is widely in favor in the US is commonly referred to as a "cap and trade" plan. It is the approach actively considered at the Copenhagen 2009 conference and the one accepted in 2009 by the US House of Representatives. The idea is to allocate an allowable level of CO_2 release to each producer (power plant, factory, or entire country), and permit them to sell, buy, or trade rights to such releases. An incentive to reduce emissions is created by cap and trade, since any decrease in emissions is directly salable on the market. The permitted allowances might be distributed free of cost or at a predetermined price. One estimate[8] has the world market for such "carbon trading" at $126 billion for the year 2008, but the market for carbon pricing has been quite volatile and it is therefore difficult to plan ahead. The use of caps at various locations implies monitoring and verification, requirements that are workable by compliance in the domestic circumstances, but one that is a major stumbling block in international negotiations. While the US has insisted on stringent verification, China has resisted any kind of international monitoring of its self-imposed target on GHG emissions.[9]

The proponents of this system like to compare it to the US 1990 Clean Air Act, which allowed coal-burning power plants with sulfur dioxide (SO_2) and nitrogen oxide (NO_x) emissions below the level of their cap to sell their credit to other utilities whose emissions were too high. It is reported that sulfur dioxide emissions in the US have been reduced some 43% since the legislation was implemented,[10] in spite of the fact that many older power plants were "grandfathered," that is allowed to circumvent the new rules.

Another aspect of the cap and trade formula provides for so-called "offsets" that would allow substitute alternatives in lieu of emission reductions. The moves creating offsets could directly reduce the carbon dioxide (CO_2) in the atmosphere, such as planting a large acreage of trees, or they could be indirect in their influence, such as policies that have the effect of avoiding deforestation. If offsets are adopted, the emissions cap for any source would be raised by any action by an amount commensurate with its effect on reducing CO_2 in the atmosphere. It remains to be decided what range of actions of commission or omission are to be included and treated as offsets.

Critics have argued that cap and trade does more harm than good,[11] that "it disables our ability to reduce greenhouse gas emissions ... [and] offers incentives for businesses that pollute." They object especially to offsets in the belief that they would be largely unverifiable. To meet some of these objections Hansen[12] has proposed to replace cap and trade with another system that he calls "fee and dividend," by which a carbon fee would be imposed on each unit of fossil fuel produced or imported, scaled in magnitude according to the amount of CO_2 that combustion of that fuel produced. The price of goods would presumably rise to reflect the amount of fuel used in their production, thereby discouraging the use of that fuel. Following Hansen's idea one step further, the collected fees would be distributed to the public as a dividend, where they could be used to purchase goods that use carbon-saving technologies. A variation on this theme has been proposed by Levine,[13] who would place the fee on the consumer good (say gasoline) instead of the original fossil fuel. He would then keep a record of the fees collected from each individual and return the sum that was collected back to each individual in the form of an account that could be drawn upon to buy goods or services that were identified as encouraging low-carbon usage.

Other plans originating in Texas, California, and Germany, and later adopted federally in the US have used "cash for clunkers" or so-called "wreck rebates" in an attempt to encourage consumers to trade in their old cars in exchange for newer, more efficient vehicles. One estimate[14] of the savings that would result from exchanging all vehicles that get 20 miles/gallon for those getting 32 miles/gallon gives an annual reduction of 1.35 billion gallons of gasoline, about 1% of total US consumption.

In the previous chapter it was emphasized that the economic pressures that can support the development of alternatives to fossil fuels are sharply dependent on the costs in any given market. The incentives are created or lost as the prices of petroleum, natural gas, and coal fluctuate. In the long run, say several decades from now, it is to be expected that prices will rise as production and reserves are diminished, but why wait for this to happen if the market pressure can be accelerated by changing fiscal policy? This fact is arguably the most important reason to support taxes on the use of fossil fuels, provided that the result is to make the energy alternatives attractive enough to foster their development. Such taxes would also produce a flow of income that could be used to support domestic research and development, to subsidize attractive new industries, and/or to fund international grants to limit emissions in developing nations. The imposition of graded taxes is in general easier to implement than cap and trade rules, which are subject to a host of subjective evaluations and political judgments. Furthermore, a tax structure could lend itself easily to rebates (in essence negative taxes) as rewards for compliance over and above the required limits.

9.3 A Complicated Question

Suppose we ask a complicated question: what must come together – to converge – if we are to break through the wall of resistance, inertia, and cognitive conservatism to develop cooperative and timely responses to the many threats (and opportunities) generated by climate change? Kingdon has addressed such a question by using the metaphor of three "streams" that must come together: problems, policies, and politics.[15] We can extend his analysis into the international arena where we are primarily concerned with how an issue on the agenda can break through the political resistance to change. The answer or answers may vary in different circumstances but there seem to be some common themes. In the first place, there must be an emerging consensus that an issue must be dealt with, that "something needs to be done about it" relatively quickly. This consensus on the need for action may arise from a crisis, a dramatic event, or a very salient change in an indicator (as with various economic "shocks"). What action to take and what policies to implement are usually more controversial because the forces resistant to change are always powerful and disagreements about how to respond are to be expected.

An example of such occurred in 2009 when the US House of Representatives passed a bill seeking to use a cap and trade mechanism to reduce GHG emissions by 83% by 2050. The law would create so-called carbon permits that could be bought and sold, allocating 35.5% of the free permits to the power sector (utilities) and only 2.25% to petroleum refiners. In response, the American Petroleum Institute organized a series of public rallies to send a "loud and clear" message to the Senate in opposition to this legislation, arguing that it would force oil companies to buy many of their permits on the open market.[16] In rebuttal, a spokesman for the research and advocacy organization called the Center for American Progress made the point that the refiners would be allowed to keep the value of the free allowances while public utilities would be required to return the value of their permits to customers. He charged that "the oil industry's goal [was] to block or weaken efforts to tackle global warming."

A second factor that facilitates convergence is the gradual development of a consensus among the experts and specialists on an issue about what they know and do not know and on what alternatives should and should not be considered. Kingdon has noted that ideas that survive in the policy debate are those that seem likely to be technically feasible, to reflect the values of the community, to not seem too costly, and to generate public support.[17] Nevertheless, even when there is a policy consensus among the experts and a problem calls for a strong policy response, very little is likely to be accomplished unless the political system (both domestically and internationally in the US case) is open to change. This can come about in a variety of ways: a change in administrations, the emergence of a strong and charismatic leader, a change in the distribution of power and ideology in deliberative bodies, a change in public opinion or public moods about an issue, or a shocking event – a 9/11, a Pearl Harbor – that undermines conventional patterns of thought and action. Consensus in the policy arena usually emerges through discussion and analysis, but a consensus in the political system usually reflects either bargaining and compromise or electoral supremacy by one party and its ideology.

The convergence of a demanding problem, a large degree of expert consensus on what to do about it, and a political system able and willing to take serious policy action creates, in Kingdon's terms, a "window of opportunity," a window that can close quickly, or not be perceived, or can be forcibly closed if the resistance to change is still powerful. Given the likelihood of opposition to change and some degree of intellectual uncertainty about the best course of action, the opportunity is not likely to be grasped without strong leadership willing to risk political capital.

There are some potential parallels here with the evolution of the debate about energy and the environment. In the first place, the problems have themselves become more compelling: scientific consensus has grown, salient events such as hurricanes and droughts seem to be occurring more frequently and to inflict higher levels of destruction, and various indicators of catastrophe (rising temperatures, melting glaciers, rising water levels, and diminishing fish stocks) seem to be growing in intensity. The policy discussions among the experts have also intensified, there is a virtual consensus on the need to act quickly and cooperatively, and many alternatives have been proposed or are under active investigation. The missing link is obvious: in a variety of ways there are political and economic obstacles to convergence, to grasping the window of opportunity that seems to be opening.

In short, we have an important degree of convergence. Energy and the environment have clearly been defined as problems that need to be dealt with by both national governments and international institutions and there is growing consensus in the scientific community on the nature of the policies that need to be adopted. And, finally, there may be more than cosmetic movement in the political arena: a new administration in Washington, aggressive plans by the United Kingdom that may spur others into action, some encouraging signs from China that it is intent on cutting its emissions and creating a competitive renewable energy industry, tentative signs that the worst of the economic crisis may be over, and growing awareness of the potential benefits of large-scale investments in renewable energy and enhanced efficiency to economic recovery now and for decades ahead. One needs to emphasize that these potentially positive developments are all at risk. For example, China is still stubbornly pursuing nationalist policies and the ongoing debt crisis in

Greece and southern Europe may generate slower growth and another recession in the world economy.

9.4 An Overall Assessment

Clearly, complete convergence between problem definition, policy consensus, and political agreement is still some way off. The negative influences may be easily listed:

1. The economic crisis is still with us.
2. Anxieties about bearing the costs of a transition to a low carbon economy are still high.
3. Businesses still fear that adopting new but costly technologies could leave them exposed to competitors who do not adopt such technologies.
4. There is continuing fear that the benefits of technological breakthroughs will be hoarded and not shared.
5. Public opinion is still not strongly supportive of the necessary policy changes (especially if they raise costs).
6. Divergent interests and perspectives still prevail.
7. Doubts are growing about whether the Obama administration's policies will match its rhetoric.
8. The developing countries are still unwilling to change existing growth strategies, to accept firm targets and timetables, or to act without large increases in foreign aid.

As one illustration of such pressures, note that while China is investing heavily in renewables, it is doing so through a strategy that is protectionist and nationalistic, largely by favoring its own domestic energy companies when contracts are awarded.[18] If other countries were to follow the same route, it could generate a trade war. China's actions in this regard may have been a response to the provision in the US cap and trade legislation that would permit tariffs to be imposed on countries that do not control emissions adequately and quickly.

On the positive side we may note that some of these problems may be diminished by moving to a narrower and more focused negotiating arena, because not every country needs to adopt every policy from the start (variable geometry, as the European Union would have it). Furthermore, there may be a demonstration effect from the actions of successful adapters, and the pace of convergence may increase, especially if newly adopted policies produce quick benefits at an acceptable cost. While progress toward cooperation obviously still hangs in the balance, the number of positive signs seems to be growing slightly faster than the negative signs. In any event, it seems more prudent to seek to deepen and extend the potentially useful trends and developments, rather than demanding massive and immediate policy changes that are neither politically nor economically viable. Besides these trends, we must acknowledge the unwelcome fact that some of the imponderables that we fear and hope to avoid may change the balance if they should show up as new indicators of looming catastrophe. In any event there is no valid excuse to delay preparations for adaptation and mitigation that could at least diminish the effects and lay the groundwork for a more rapid recovery.

9.5 Choices and Priorities

Many analysts have listed the components of a policy package for the United States to deal with climate change and energy security.[19] The issues have been on the table since the oil crisis of the 1970s and the overheated debate about the limits to growth. However, what has been missing from the discussion of most of these lists is an analysis of priorities – what should be done, when, and why – and a sense of how to deal with the constraints on policymaking in a very difficult political and economic context. In the matter of priorities we hope to point in some useful directions by dividing the field into segments according to the time scale appropriate to each choice. Regarding policy, we recognize that the conventional debate between advocates of a free market approach and advocates of the need for strong government intervention has always complicated the problem of choice. From our perspective, it is very difficult to see how much can be accomplished without government subsidies for renewables and for research and development of a number of other technologies, without government setting a cost for carbon emissions, and without government creation and support of a legal framework for private investment. But we should also note the obvious: choosing one approach to the exclusion of all the others is not sensible or prudent. Private investment and private initiatives are also an important component of an effective response.

To overcome governmental inertia and the tendency to procrastinate there is demand in some circles for quick and forceful action which may not be politically feasible. It is more important to choose policies that are efficient, effective, and equitable to a wide range of economic groups. The contextual constraints also suggest that it would be prudent to choose short-term policies that are likely to generate relatively quick returns for the initially skeptical public at large and that do not require vast expenditures and heavier tax burdens to fund speculative projects. Above all the policies must be flexible enough to respond quickly to either positive or negative signals and able to be expanded if the need or the opportunity arises. One also does not want to ask weak and heavily burdened governments to take actions that threaten political defeat or a significant loss of public support. There is a politically prudent rule here that is crucial: do not ask too much of already burdened governments and do not expect them to take big risks or to be completely transparent about costs or unanticipated consequences.

We have not yet spoken about the economic dimensions of the problems created by global warming either now or in various future periods. The economic crisis that erupted in 2008 has obviously put sharp constraints on the money that is likely to be available; it may even be difficult to maintain current spending levels for investments in renewable energy and foreign aid if growth does not resume and optimism about the future is not restored. Perhaps most significantly, the interaction of so many variables in so many different contexts and time periods suggests that any set of numbers is largely notional and may imply a spurious degree of precision. Thus we agree with Yvo de Boer, the executive secretary of the United Nations (UN) Framework Convention on Climate Change, when he declared recently that estimates of cost "remain a moving target. Starting to plug the hole right now is more important than determining its exact future size."[20] In this connection it should also be noted that Nobel economist Paul Krugman believes that "claims of immense economic damage from climate legislation are … bogus … [and] that the best available economic analyses suggest that even deep cuts in greenhouse gas emissions would impose only modest costs on the average family."[21]

Despite these uncertainties, some sense of the magnitudes involved in adapting to climate change is useful. This is especially the case because many who are opposed to the need to diminish global warming as rapidly as possible rest their case on cost estimates that are analytically unsound and seemingly designed to terrify the public rather than educate it. No one, of course, denies that there will be adjustment costs in establishing a new energy economy or in adapting to the existing direct and indirect costs of climate change but those costs should be bearable. Indeed, if some of the scientific forecasts of the likelihood of imminent (10–20 years) catastrophes are accurate, the costs of the failure to act now could dwarf the costs of acting prudently now and in the next two decades.

For the developed countries, estimates of average costs are especially difficult because each country is likely to have a very different mix of renewable and traditional sources of energy, differently priced. In addition, it is very difficult to know what items should or should not be included in the estimates. Barring a sharp and sustained price rise for fossil fuels, renewables are unlikely to replace very large segments of oil and natural gas use for several decades to come, and even a process of partial replacement will require government subsidies for some years. Nevertheless, while the opportunity costs of spending on the creation of a new energy economy may be high, most developed countries can benefit significantly from the contribution of such spending to the current stimulus packages. In any case, as with wartime spending, there is no sensible alternative to finding the necessary resources.

Paul Krugman has summarized the best currently available estimates of costs for the United States. He notes that a recent study by the non-partisan Congressional Budget Office has concluded that the recently passed Waxman–Markey energy bill "would cost the average family only $160 a year, or 0.2 percent of income" in 2020. The burden would rise to 1.2% of income by 2050, but since real gross domestic product (GDP) would be about 2.5 times larger, the "cost of climate protection would barely make a dent in that growth."[22] And, of course, these estimates of costs do not include the obvious benefits of limiting the worst effects of global warming. The increasing polarization of the debate on climate change in the United States risks generating a kind of Gresham's Law of politics, with partisan posturing driving out sensible analysis; efforts thus need to be made constantly to resist ideological, self-interested exaggerations of potential costs.

An even more vexing economic problem concerns the developing countries. In fact, their demands for a substantial commitment of foreign aid as compensation for damages inflicted by developed country emissions now and in past decades, and to deal with adaptations to climate change, became one of the major obstacles to an agreement at Copenhagen. Early estimates had suggested an annual figure of about $100 billion for 10 years, above and beyond existing levels of foreign aid. More recent and more technically sophisticated analyses have sharply increased the anticipated amounts necessary for successful adaptation. Because of previous development failures, especially the failure to develop functioning infrastructures, one recent study has suggested a need for $315 billion per year for 20 years to remove the infrastructure deficit and another $16–63 billion per year to upgrade infrastructure to deal with climate change.[23] A new UN report has suggested an annual figure of $500–600 billion for 10 years to allow developing countries to grow without reliance on "dirty" fuels.[24] Although less than the total amount committed under current stimulus packages, the magnitudes are daunting even as rough estimates. Put on top of the stimulus expenditures, it does not seem likely that such sums are either politically or economically feasible. It appears, however, that some kind of financial commitment, presumably between $100 and $600 billion per year, will emerge from the negotiations, if only because there is virtually no hope of slowing global warming without strong commitments to a new energy paradigm by the developing countries. Such spending is so monumental in current circumstances that it may increase the pressure for greater spending on research and development of renewable sources and/or some of the various climate engineering proposals that have been put forth with the object of reducing costs markedly. In assessing the impact of these figures, it should be added in passing that if all developed countries committed to the goal of providing 0.7% of GNP in official development assistance, the sum would come close to meeting the $300 billion target. To date not all have made this commitment, nor has the US. In evaluating this reluctance, it must be admitted that the effects of foreign aid have often been disappointing: too much of it has been wasted on the wrong projects or lost to corruption or incompetence.

The poorest countries and the middle income developing countries already generate about half of the total carbon emissions and their share is likely to grow steadily. Continued deforestation, the growth in power plants using "dirty" fuels, and a continued commitment to the standard growth strategy of rapid industrialization are by themselves a recipe for disaster. Aside from any severe costs in terms of natural disasters, declining health, increased conflict, food shortages, and massive population movements, the most recent World Bank study[25] suggests the costs to Africa could come to 4% of GDP and to 5% in India. This staggering cost indicates that the developing countries have more at stake in the Copenhagen negotiations than any other group of countries, implying that there is some potential for bearable compromises by all sides at the conference itself.

What may make these compromises difficult to achieve is not merely the magnitude of Third World financial demands – which obviously dwarf current aid levels at a moment in time when the effectiveness of foreign aid is being sharply challenged and there are many alternate calls on available resources – but also that the demands reflect deeply felt grievances and resentments. The poor feel that the developed countries created the problem and that it should not and cannot be resolved by demanding sacrifices in future growth by those who can least afford it. Moreover, since the poor countries view aid as compensation for past injustices, they do not want any restrictions on the aid that they do receive.[26] In effect, they refuse as yet to accept the standard Western demands for accountability and transparency for all aid disbursements. Western doubts about the wisdom of this have been exacerbated by the failures of many aid projects and by the weaknesses of the developing countries in effective implementation (an issue to which we shall return). In any case, some compromise will presumably be negotiated eventually because it is clearly in the interest of all parties committed to reducing global warming, but where that compromise will fall is a matter of guesswork at this time. If there is no compromise or if an agreement is perceived to be largely cosmetic and ineffective, demands for increased expenditure on geo-engineering projects will surely escalate, an issue to which we shall return later in this chapter.

One last comment on this issue is potentially significant. The bulk of current Third World contributions to global warming obviously come from a small group of countries: China, India, Brazil, and Indonesia (the last two because of deforestation). There are some signs that these countries are increasingly becoming aware of the dangers to themselves of global warming and the possible benefits of attempting to profit from technological breakthroughs. They were also obviously intent on maintaining bargaining leverage before, during and after the Copenhagen conference. Their hard-line stances have persisted, however, and at a time when the ability of the developed countries to respond generously is rather strikingly limited.

The hard-line attitude has not come from indifference or ignorance. China, in particular, although unwilling to accept firm targets and timetables, is making a major effort to reduce its emissions and to join the effort to profit from the development of alternative fuels. Also, Brazil and Indonesia have been sending signals recently of a willingness to sharply cut deforestation (which accounts for an important share of total emissions), if they are sufficiently compensated. None of these countries want to be blamed for a self-interested failure to cut emissions, none want to miss out on possible economic gains, and none want to be devastated by the consequences of global warming.

A (very) cautious optimist might see in this, if through a glass darkly, tacit signs of a growing awareness that self-interest and the global interest are linked, as are development and the effort to diminish global warming. But two sides talking by each other and not trying to understand what the other needs or wants is not a recipe for negotiating success. At the moment, there is no way of knowing whether this gap will persist, diminish, or increase. The delay in moving toward compromise may be increasingly costly.

9.6 Caveats

The choice of policies in an arena that is so complex and uncertain is rather like a permanent juggling act. Each of the balls – scientific, political, economic, and the unanticipated – bounces to its own rhythms but can also be deflected by sudden, sharp movement in the others. For example, focusing on the familiar "low hanging fruit" (say, conservation or increased efficiency of use) makes eminent practical sense in the short run but it could also prove inadequate if global warming continues to increase and other countries lag in their responses. If such occurs, we should be prepared to act vigorously to develop public support or, if the need for emergency action is imperative, acting without such support. In general, decisions should seek to build time and support for more aggressive policies in the future, if and when they are needed.

Early funding for research can be justified if the chosen projects are a form of long-run insurance. Steven Chu, Energy Secretary in the Obama administration, has made a number of important proposals for research and development, most of which should receive funding now although it may be decades before they can be realized. As noted earlier, the long run may become the short term if conditions deteriorate. Chu's suggestions include painting the roofs of buildings and road surfaces white to reflect sunlight back into space, building a "smart" electricity grid, and burning nuclear waste in special reactors that will transmute it into more benign elements. Chu has also cut funding for projects that he considers ill-advised.[27]

For longer term decisions, much depends on the successes of short-term policies, as well as further developments in global warming, energy security, and unanticipated external developments (economic "shocks," political upheavals in key countries, etc.). Heavy spending on research and development will be appropriate for all three of the time periods that we shall discuss, but a sharp distinction between different time periods is somewhat artificial, because some medium-term policies (say, building nuclear reactors) and some long-term policies (say, nuclear fusion or some geo-engineering projects) require short-run and sustained funding. Initial spending should be spread out among a reasonable number of possibilities, but as one or another of these possibilities seems more promising, spending should be both more concentrated and more extensive. It is short-sighted to gamble early on any large, particular project.

The international arena is equally important but the obstacles to cooperative agreements are likely to be even more severe than in the domestic arena. Divergent interpretations of national interests, different levels of development, different political schedules, and different values and ideologies complicate the bargaining process. In the circumstances, if we assume that speed is as important as efficiency, effectiveness, and equity in international policy responses, it makes eminent sense to concentrate initially on cooperative and *substantive* (not rhetorical) agreements among the current major polluters. One hopes that the latter will constitute a "coalition of the willing;" if not, various carrots and sticks may have to be used to "induce" cooperation. Global conferences like Rio and Copenhagen will become more productive if short-run cooperation among the major polluters has been reasonably successful, and if new technologies and additional foreign aid become available. It is to be hoped that new technology would be shared and that fears of competitive disadvantages would not generate excessive nationalism and protectionism. This is one strong reason to favor international agreements that set standards for the system as a whole. Finally, it is worth re-emphasizing again a point made earlier: international cooperation can only succeed if it rests on strong national policies to reduce emissions and move toward a new energy economy.

9.7 A To-Do List

Before discussing specific policy choices for different time periods, it may be useful to summarize the characteristics that we hope will or should be present in the policies we choose. It may also be useful to note that we advocate a two-track strategy that covers three time periods and also includes a possible third track if conditions deteriorate sharply. In sum, we shall suggest appropriate strategies to pursue appropriate policies in different time periods.

We shall seek policies that adhere to well-defined guidelines, as follows:

1. Goals should be set that are likely to be achievable in one or another time period; lead times to develop and implement particular policies must be very clear over the short term, moderately clear for the intermediate period, and may be more speculative for the long run.
2. Costs must seem bearable to both the policy community and the public, although "bearable" in this context must also be adjusted to the perceived scale of the threat at any particular moment.
3. Potential side effects must always be taken into account; initially, very risky projects should be tested and evaluated experimentally before implementation.
4. Policies must be politically acceptable to a sufficient majority, both domestically and internationally, unless the threat of global warming has become so grave that immediate action is imperative.
5. Opting for any single policy approach is dangerously premature and flexibility of funding and support should be maintained.
6. Policies cannot be inequitable, benefiting one segment of society over another and, if such an outcome seems likely, compensatory schemes must be established.

These ideal requirements provide useful benchmarks for the specific choices that follow.

Regardless of what happened at Copenhagen, we have a moral and practical responsibility to do as much as we can as quickly as we can to deal with the immediate effects of global warming and to prepare to deal as best we can with its long-term effects. This is our first track and it encompasses policy actions for the near future, the medium time period, and the long run that fulfill as many of the desired policy characteristics as possible. We seek the best merger of the feasible and the necessary that we can achieve. Our preference is for doing this both nationally and internationally by coalitions of the willing that seek to overcome the deficiencies of the policymaking process by a process of gradually accelerating incrementalism.

We must also begin to prepare for a shortfall of track one by increasing expenditures for research and development of some untested technologies that are based on recognized science, as well as for a variety of unconventional and novel geo-engineering technologies with as yet unknown consequences. This is track two, which should be pursued simultaneously with track one. A worst case scenario: suppose that neither track one nor track two suffices and global warming accelerates with all its attendant environmental, political, and social dangers? Acknowledging this possibility, planning staff in relevant governmental and international organizations should be devoting some intellectual energies to what sanctions and/or penalties might be necessary in such dire circumstances.

Notes and References

1. D. S. Battisti and R. L. Naylor, "Historical Warnings of Future Food Insecurity with Unprecedented Seasonal Heat," *Science*, Vol. 323, January 9, 2009, p. 240.
2. "Labour Orders Green Energy Revolution," *The Guardian*, July 16, 2009, pp. 1, 6–7, and 32.
3. Ibid.
4. "A Good Climate for Development," *The Economist*, August 6, 2009, p. 9.
5. David Adam, "Catastrophic Warming 'in our lifetimes' – Met Office," *The Guardian*, September 28, 2009, p. 1, and "Questioning the Invisible Hand," *The Economist*, October 17, 2009, p. 66.
6. *Danish Climate and Energy Policy* (Copenhagen: Danish Energy Agency, 2009) and *Renewable Energy Policy Review* (Copenhagen: Danish Energy Agency, 2008). We have relied on these reports for our comments.
7. Quoted in Thomas L. Friedman, "The Copenhagen That Matters," *New York Times*, December 23, 2009, p. A27.
8. K. Capoor and P. Ambrosi, *State and Trends of the Carbon Market 2009* (Washington, DC: World Bank, 2009).
9. John M. Broder and James Kanter, "China and U.S. Hit Strident Impasse at Climate Talks," *New York Times*, December 15, 2009, p. 1.
10. James Hansen, "Cap and Fade," *New York Times*, December 7, 2009, p. A27.
11. Laurie Williams and Allen Zabel, "Cap and Trade does More Harm than Good," *Philadelphia Inquirer*, June 24, 2009, p. A19.
12. James Hansen, "Cap and Fade," op. cit.
13. Marshall Levine, personal communication, December 1, 2009.
14. Lisa Margonelli, "Let's get Serious about Auto Sales, Eco-incentives," *New York Times*, May 16, 2009.
15. John W. Kingdon, *Agendas, Alternatives, and Public Policies* (Boston: Little, Brown and Company, 1984).

16. Clifford Krauss and Jad Mouawad, "Oil Industry Backs Protests of Emissions Bill", *New York Times*, August 19, 2009, p. B1.
17. John W. Kingdon, *Agendas, Alternatives, and Public Policies*, op. cit., p. 126ff.
18. Keith Bradsher, "In China, a Shield Goes Up for Energy Firms," *International Herald Tribune*, July 15, 2009, p. 13.
19. See, for example, various essays in Kurt M. Campbell and Jonathan Price, eds, *The Global Politics of Energy* (Washington, DC: Aspen Institute, 2008).
20. Quoted in Tom Zeller, Jr., "A High Cost to Deal with Climate Shift," *International Herald Tribune*, August 31, 2009, p. 18.
21. Paul Krugman, "The Truth: It's Easy Being Green," *New York Times*, September 25, 2009, p. A25.
22. Ibid.
23. Martin Parry, Nigel Arnel, Pam Berry, *et al.*, *Assessing the Costs of Adaptation to Climate Change: a Review of the UNFCCC and Other Recent Estimates* (London: IIED and the Grantham Institute, Imperial College, 2009).
24. Neil MacFarquhar, "A Sobering Estimate for Clean Global Energy," *International Herald Tribune*, August 10, p. 12.
25. *World Development Report 2010: Development and Climate Change* (Washington, DC: World Bank, 2009), pp. 12 and 45–6.
26. See "A Bad Climate for Development," *The Economist*, September 19, 2009, p. 77.
27. See "The Alternative Choice," *The Economist*, July 4, 2009, p. 62.

10

A List of Priorities

In this chapter we provide a framework that may be used to answer several linked questions: what policies do we advocate for which actors in what time period? The *why* question has already been answered: to devise policies that can slow climate change by helping to create a new, low carbon-based economy, and to do so in a way that is efficient, effective, equitable, and timely. The *how* question may be the most difficult of all, as the last chapter emphasized. Much will depend not only on effective political leadership but also on the evolution of the threat of global warming itself. We shall here draw on material presented in prior chapters to make some suggestions on how to improve the prospects for policy agreements. The net results of our analysis are presented as a summary listing in Table 10.1 at the end of the chapter, in which various suggested pursuits are divided according to whether they are items of implementation or further research. Some of the proposed

The Challenge of Climate Change: Which Way Now? 1st edition.
By Daniel D. Perlmutter and Robert L. Rothstein.

changes are domestic, others call for international agreements. All the proposals are set out in the form of a timetable to show priorities according to their degree of difficulty, either because of scientific, technological, or political uncertainty.

10.1 Short-Term Gains: Less than 10 Years

The policies in this section have a number of characteristics that ought to be noted. In the first place, they can all be implemented nationally or even subnationally without waiting for international or even regional agreements. Nor do they require strong sanctions for non-compliance. Needless to say, their cumulative impact will be much greater if other states adopt similar policies as a kind of uncoordinated but reinforcing incremental response to global warming. Secondly, the scientific and technological bases for such policies are reasonably well known, although some will obviously benefit from continued research. Finally, some are already providing efficient and effective benefits on a relatively small scale, others are more costly and will require time for implementation, but none as yet seem to raise insuperable problems in terms of equity. Perhaps above all, they are politically and economically feasible in the US, especially if the Obama administration is willing (or compelled) to expend some political capital. Enhanced research spending on other options is also imperative, but should be bearable in present circumstances. For example, the US Department of Energy Secretary Steven Chu already has at his disposal an annual budget of some $26 billion as well as $39 billion of the recent stimulus package. In short, there is no reason not to act now: we know what we must start doing, we can expect reasonably quick returns on such investments, and they are clearly in the national and international interests.

Energy conservation programs have already begun to yield benefits in some countries, as consumers have switched to more efficient devices and appliances, including for example fluorescent light bulbs, heat-shielding windows, and better insulated refrigerators. Savings in industrial and municipal applications are available by redirecting what would otherwise be waste heat; that is, excess heat left over from some higher temperature operation but still warm enough to be useful for domestic space heating.

We have already made reference to the variety of US fiscal policy steps that provide incentives to encourage individuals to act in the public interest.

These include rebates for turning in automobile "clunkers," tax credits for installed insulation, for windmills, and for solar collectors. The cap and trade legislation is designed to reduce greenhouse emissions from industrial plants.

Among the gains that are achievable in the short term, those for which the science is well understood and for which the required technology is readily available, are improvements in transportation vehicles and systems. The increases in CAFÉ requirements in the US that were mandated in 2009 are modest, and there could be significant reductions in the use of petroleum if these standards were increased again to at least the European levels, if not to an even bolder standard. Additionally, along this line, transitions to a greater use of diesel engines could further emulate the gains already known in European autos. And finally, there is great advantage to be sought in a public bus and/or train service that is clean, convenient, and on a frequent enough schedule to attract those who would otherwise be drivers in their own cars.[1]

Also easily and immediately available is solar thermal energy. This source is most readily transformed into home and shop hot water and space heating and could as well be used in industrial settings where extremely high temperatures are not required. Its application to high temperature processes does not need any new scientific breakthrough but could benefit from some engineering improvements, whether in concentrating the solar radiation or in storing and recovering the intermittent energy.

Next in line for ease of exploitation are the proven methods for collecting wind energy. The use of large propeller-driven generators of electricity are scattered over Europe and parts of the US, the electricity is already fed into large area power grids, and the system is supported by popular acclaim and encouraged by governmental grants. Location of windmills for power are limited in some areas by the high cost of installing transmission lines, but this will be a less determining factor as other fossil fuel energy costs rise.

Direct use of solar energy via photovoltaic cells is presently easily available but relatively expensive. This has led to installations in a variety of special applications, mostly in places that are remote or not easily serviced, for which capital investment is not a primary consideration. Major changes in this situation could well materialize during the next decade; they will require new technical developments in the direction of less expensive materials of construction or increased efficiency in transformation of sunlight into electricity.

The better performance from batteries that has emerged in recent years was driven initially by a need for power in small electronic devices and then by automotive requirements. The improvements are impressive and it appears that an all-electric car will be available within a few years, even though the driving range between chargings of the battery is still a sharp limitation. Looking further ahead, consumers could more easily accept the shorter range of a plug-in electric vehicle if their commuting distances were shorter. This in turn would require a significant shift in living and/or working locations, changes which have already occurred to some who work at home at least part of the week. To be of help in handling the issues of global warming it will be necessary to use renewable energy as a source of the electrical input to the battery. This is not yet available on a large scale, but there is reason to expect this capacity to increase in the near future.

To the extent that emphasis is placed on currently available technology for making liquid fuel from hydrocarbon sources, the Fisher–Tropsch (FT) process must be high on the list. While it was originally developed with coal in mind as the energy source, this starting material does not help in reduction of carbon dioxide (CO_2) emissions. Instead, to maintain a balance of greenhouse gases in the atmosphere the route to the required feed hydrogen and carbon monoxide needs to be an agricultural product, preferably one that is not edible. This combination would be designed to recycle the CO_2 produced by the burning of the liquid fuel by absorbing the gas in the growth of the plant source. The same advantage could be claimed for biobutanol, provided that the fermentation mash was an acceptable plant product. To serve our purposes, it could well take a decade to develop better plant input sources for either of these processes.

Finally, in this catalog for a first decade, there is the possibility of building reservoirs for pumped water or compressed air. The technology for either or both of these is fully developed and available and can use any source of power that needs to be stored. One shortcoming of compressed air stems from the variation in pressure as the stored energy is recovered. Whereas batteries, for example, provide constant voltage over their entire discharge, pressure changes in air storage call for engines that can deliver power at a constant level as pressure drops in the storage container. In compensation for this weakness, compressed air can transfer power at very high rates to effect rapid acceleration. Other advantages of compressed air over battery energy storage are the longer life of pressure vessels and the low toxicity of the materials used. Again, high investment costs are the only important obstacle to this technology, and its use will depend on what alternatives exist at a future time when it is more essential to store energy than it is today.

While it is not directly a process for energy production, the development of new nuclear power stations reopens questions on the storage of nuclear wastes. At this time in the US the wastes are stored in 131 locations spread over the country, a system that raises serious doubts with respect to security and efficiency. The Yucca Mountain site in Nevada was chosen to consolidate the storage, but although some $9 billion was spent to establish that the site was geologically acceptable, its use was ultimately stopped by political opposition. An option being considered now is to recycle unused fuel for use in new advanced reactors.[2]

The earthquake hazards that have been associated with deep drilling for enhanced geothermal systems were described in Chapter 3. Here our concerns for the next decade should include geological research to establish the safe limits on such operations. We need to know how deep we can drill in what kinds of underground deposits, how close we can come to major fault lines, and whether dangers are associated with porous materials as well as with those that must be fractured with high pressure injection. In a somewhat related geological issue, questions are being raised about possible groundwater pollution in the wake of natural gas production via hydraulic fracturing of underground shale rocks, and the US Environmental Protection Agency (EPA) is reviewing what is known to determine whether drinking water in some locations has been contaminated by the chemical additives used in the fracturing process[3].

In addition to this list of technical goals for a decade ahead, attention should be paid to ways of creating political "facts on the ground" that would make possible the necessary political support. A process of change could then begin that could have very profound effects both immediately and over time. Recognizing that governments and elected representatives are not likely to override the interests of their financial supporters unless they feel strong pressure from their grassroots constituents, we might ask what developments could serve to break down the wall of inertia, the lack of deeper engagement among the voting population that has led to a confusing debate between experts and advocates on both sides.

Because it is the young, especially those of college age, who have the most at stake in this debate and who have the greatest opportunity to learn about it from the large number of environmental courses that have spread across the educational universe, it is this group of citizens that should be most effective and committed. Just as protest movements against the Vietnam War and for civil rights spread quickly across the campuses, and ultimately to

Washington, an environmental movement is most likely to emanate from campus activism that spreads and deepens.

If sizable, such a movement would certainly grab the attention of policy-makers in Washington and other political arenas. Political entrepreneurs will certainly see a window of opportunity opening and seek to build on it and spur it along. In short, we advocate the creation of groups whose members seek to persuade their institutions to make a significant commitment to reduction of their carbon footprint. The benefits of personal commitment would be found in a variety of small and familiar measures: not leaving computers on standby, buying more efficient appliances, or using bicycles for transportation, but the spillover to greater political pressure might have a more substantial effect. While quantifying the potential effects of this movement is premature, we believe that the cumulative results could be massive.

10.2 Medium-Term Improvements: 10–20 Years

Some of the approaches to energy conversion have in common that the necessary scientific underpinnings are known but the engineering difficulties are big enough to demand considerable time for application. For such applications we should plan on efforts that demand more than a decade to design, build, and complete. A prime example for such a time scale is a nuclear power reactor, for which planning, designing, and construction require a period of about 10 years. This interval could be shortened if near-term negative developments should introduce a sense of urgency. Looking further ahead, there is a need to establish a standardized nuclear reactor design for safety and efficiency in any future expansion of this industry. New designs have been proposed; one is based on a set of smaller modules, each of about 45 MW capacity.[4] Safer and cheaper due to the smaller size of each unit and their underground placement, the combination of 10 or 20 such reactors would equal the capacity of current nuclear installations. The design is currently in the process of being evaluated by the US Nuclear Regulatory Commission.

Considerations of what it takes to extract power from waves and tides show that the science of wave energy is well known, as is also the knowledge of tides that react to the gravitational attraction of the moon (and sun). The engineering details of the devices that would collect this energy still need to be developed and tested empirically in place. Optimal locations are yet to be identified. Furthermore, even when we know that barrages and tidal barriers can be built, and where and how to build them, the time for such constructions would cover many years. As noted before, Britain has chosen to build a barrage on the Severn River estuary at a projected cost of $29 billion to be funded over a period of 20 years.

Realistic plans for better batteries and fuel cells will also need extended times for research and development. The two both involve chemical conversion, the battery to transform electricity to a reversible chemical storage, the fuel cell to transform energy already available in chemical form into electricity. A battery is an energy storage device; a fuel cell is an engine, a user of fuel. Nevertheless, they are linked by the need to find improved catalytic surfaces for their electrodes (anode and cathode). The use of batteries in hybrid and plug-in automobiles is the most visible large-size and large-scale application. Although all the major auto makers have promised to have plug-in vehicles in the market in the next several years, a 2009 study by the US National Research Council concluded[5] that it will be decades before enough such cars are on the road to significantly affect our petroleum use or carbon emissions. The barrier to large-scale acceptance is the potential cost of the battery pack, estimated to be about $14,000. With auxiliary switching and control devices, this would add a cost of perhaps $18,000 to the purchase, and price many small cars out of the mass market. The incentive for an improved battery remains very great and ongoing further research now could be very well rewarded a decade later.

While fuel cells can in principle use any hydrocarbon as an energy source, and demonstrations of fuel cells based on oxidation of methyl alcohol have also been demonstrated, research on practical applications have been exclusively focused on hydrogen as a fuel, because if new fuel cells were designed or modified to be able to process more conventional fuels such as natural gas or some petroleum fraction, it would produce CO_2 and lose its claim to be carbon neutral. Since dependence on hydrogen sources suffers from all the drawbacks already detailed above, advocates of large-scale fuel cell use rest much of their hope on cheaper hydrogen to be obtained from water splitting. To balance the picture, it should be added that fuel cells have an inherent advantage in efficiency over all furnace combustion regardless of the fuel that is oxidized. This means in effect that less fuel is used and less CO_2 released for a given amount of energy that is converted into electric current.

As noted before, the development of enhanced geothermal systems that can extract energy from deep deposits under the surface depend on a fuller understanding of any attendant earthquake hazards. Assuming that the needed scientific information is gathered during the first decade of investigation, practical installations should be built during the following decade. The extent of such efforts must remain uncertain until the earlier stage of this development is completed.

10.3 Long-Term Solutions: More than 20 Years

Many of the technological ideas that have been advanced with the goal of controlling global warming are based on science not yet fully developed or on engineering never tested. Their promise may well turn out to be entirely fulfilled, but it is crucial to recognize that even with the best intentions time will be needed before they will be used for practical climate control.

Important examples among such proposals arise from the very desirable goal of sequestration. Alas, to date only a single method of trapping CO_2 can be claimed as likely to work: that of confining the gas underground in a geological formation, such as a depleted oil well. Having never been used for an extended test, this method too is still open to some doubt until it is demonstrated. A variation on this idea is being tested for the first time by trapping CO_2 emitted from a functioning power plant. The American Electric Power company started in 2009 to inject CO_2 into a very deep porous layer (7,800 feet (2375 m) below the surface) adjacent to one of their coal-burning power plants. The operation is expected to use a large fraction of the power that the plant generates, perhaps some 15–20%, but if successful in all other aspects, this could be a first demonstration of the claims put forward for what has been called "clean coal." It remains to be seen whether the captured gas will remain in place, and whether it will have harmful effects as it displaces underground water.[6]

All other methods of storage for the huge quantities of CO_2 that would need to be held are at this stage largely conjecture. Although their scientific bases are undeveloped and they must today be classed as entirely unproved, the ideas are worthy of further pursuit, offering fertile ground for a decade of research on new approaches. From geologists, we need to know where deposits of basalt rocks are easily accessible, and under what conditions of temperature and pressure they are to be found. From chemists we should ask how we might cause the oxides in basalt rock to react with CO_2 to form stable underground carbonates. And from both biologists and chemists, we will want to ask how long carbonate and bicarbonate salts dissolved in the ocean will remain in stable forms, and how such changed conditions can affect marine life.

If and when the time comes that a proven method for sequestering CO_2 becomes available, all alternatives for energy transformation that concentrate greenhouse gases at point sources will immediately become more attractive. This would mean, for example, that electricity-generating power plants, whether fueled by coal, gas, or oil, would not contribute to the CO_2 burden in the atmosphere. As a result the millions of cars that currently produce CO_2 could be replaced by all-electric vehicles freed from greenhouse gas emissions.

Cheap power obtained by nuclear fusion (as apposed to uranium fission) has been a recurrent dream for more than half a century. It has always been out of reach because the engineering difficulties are prodigious. Recent research has opened some new avenues in this effort, but we are still far removed from practical results. Judging from the record, we should not expect significant results within the next two decades, but can include them in our long-run hopes.

It has long been known that its constituents, hydrogen and oxygen, could be obtained from water by electrolysis, but practical application of this process has been prevented by its very low efficiency and the relatively high cost of electricity. A very high temperature alternative is also being investigated[7] as a means of splitting water; it uses a reflective surface to concentrate sunlight on a catalytic chemical reactor, and avoids the need for electrolysis entirely. Yet another avenue to water splitting has arisen from recent research: in this approach sunlight captured by an organic dye initiates a series of steps that mimic aspects of the chemistry that occurs in plant photosynthesis. In one version of this process, which is being worked on at Rutgers University,[8] a catalyst-impregnated membrane is integrated into a dye-sensitized solar cell (DSSC), and produces the off-gases hydrogen and oxygen via intermediate titanium dioxide nanoparticles. So far the yields from this approach have been too small to justify the use of the expensive catalyst materials that would be required in practical devices, but further work on improving this process is in progress in a number of academic laboratories. Any of these alternative approaches could, down the line, offer great benefits, but it must be admitted that their realizations are still far off.

On the other hand, we have good reason to believe that we will learn during the next decade how to extract liquid fuels from plankton grown in ponds or chemical reactor vessels. Success in this enterprise will raise questions as to whether similar photosynthetic manufacture should be attempted on a larger scale by fertilizing and seeding the oceans. The fears of possible unintended consequences following such a move should be sufficient to stop it, unless the data gathered in the initial phases of this work give good reasons to proceed. Going forward with such a step would have international consequences and simple caution should call for international agreements where possible.

Given this recent history and all the obstacles to international agreement on the response to global warming, it is prudent to also have a plan B, that is, an approach that could at least be a fall-back position. What options would be left to us if after a prolonged period of doing business as usual we find that the dire predictions have become reality? Is there any hope of closing the barn door to save the colts, even after the parent horses are gone? To answer such questions we need to examine recourses in a world in which global warming has already occurred, severe enough to cause serious damage to human life via floods, droughts, storms, and the associated social and economic changes. Having failed by political compromise to prevent large-scale release of greenhouse gas emissions into the atmosphere, we will then be increasingly dependent on a scientific–technical fix to reverse the processes created by our own folly. Several such geo-engineering remedies have been proposed, also called *climate engineering*, but it must be admitted that besides the questions on feasibility that arise from the technical uncertainty attached to each of the proposed choices, there are major concerns about possible unforeseen, unintended, and perhaps irreversible consequences of world-wide actions affecting our entire planet.[9]

One geo-engineering idea stems from the observation that fine particles suspended at high altitudes reflect back into space a large part of the incoming solar radiation. This phenomenon was early associated with large volcanic eruptions that spew dust high into the air. Such an event in 1815 in Indonesia was huge enough to produce a "year without a summer" as far away as North America and Europe. More recently, data collected months after the 1991 eruption of the Mount Pinatubo volcano showed substantially reduced temperatures over a very large area, an effect that lasted for several years. With these changes in mind, it has been suggested that greenhouse-induced warming of our planet could be reduced by releasing sufficient reflecting (sulfate) particles into the stratosphere.[10] Techniques with this objective in mind are classed as SRM, an abbreviation for *solar radiation management*. In cautioning that temperature is not the only important measure of change, Hegerl and Solomon[11] have pointed out that such an action is also connected with considerable risks and could be a dangerous level of interference with the climate system. If the Pinatubo data are examined, for example, it is found that precipitation was reduced during the cooling period, an effect that if general over a prolonged period would be expected to cause major droughts, conflicts over water resources, and political instabilities over large parts of the globe.

A second SRM approach to climate engineering is based on cloud modification at lower altitudes. Noting that white clouds form in the exhausts from ship engines, the proponents of this idea advocate creating an extremely fine mist of seawater droplets that would join and alter existing clouds in order to whiten them and increase their solar reflectivity. In addition to the cost and engineering problems connected with spreading effective particles over a large ocean area, there remain the uncertainties of possible secondary effects on upwind and downwind land masses: excessive or inadequate rainfall leading to floods or drought. This approach does have the advantage of short time responses and reversibility; it could be stopped on short notice if it were found to be ineffective or seriously damaging in any way. On the other hand, it has been pointed out[12] that any change that limits temperature rise without removing CO_2 from the atmosphere leaves unchanged the effects of acidification of the oceans on marine life.

Another geo-engineering idea is an attempt to remove CO_2 from the atmosphere by photosynthesis in growing ocean plants, particularly plankton. This would call for a program of fertilizing the ocean by adding salts of iron, nitrogen, and phosphorus. The idea is to produce algal blooms that would absorb CO_2 from the air by photosynthesis and then sink to the ocean floor. To distinguish such geo-engineering approaches from SRM they are labeled as CDR for *carbon dioxide removal*. To date only small-scale fertilization experiments have been run (since 1993) with the object of stimulating the growth of plankton or algae in the open ocean. An optimistic assessment estimates that a scaled-up treatment might remove as much as a billion tons of CO_2 annually. There are on the other hand fears of irreversible, unintended consequences from such a treatment of the oceans, including local disruption of marine ecosystems and/or emissions of the potent greenhouse gas nitrous oxide. With this in mind, the London Convention Treaty created under the auspices of the United Nations has approved only limited scientific study rather than widespread use of this technique.[13] It must be said that once such a program has been followed on a large scale, it would be virtually impossible to stop it, since the fertilizers would remain in place in the oceans for a very long period.

As a political question, climate engineering has one feature that is different from any of the many international agreements that have been sought: its implementation does not require unanimous agreement or steadfast enforcement throughout the world. This appearance of ease of unilateral actions that do not require treaties with other sovereignties might suggest an advantage, but it could also result in angry and/or hostile responses from nations that perceive that they have been damaged.

10.4 Plan A and Plan B, Simultaneously

There is, of course, a potentially wide gap between statements of good intentions containing a long list of specific plans and what actually transpires in the enactment and implementation processes. A potential gap is hardly surprising: ambitious policy proposals are rarely passed in total, some interest groups will argue that the proposals go too far (i.e., their interests are threatened), others that they do not go far enough (as some environmentalists have already noted), and those actually responsible for implementation may have their own views about what should or can be done or may be less committed or less competent.

A recent Op-Ed piece by two experienced environmental negotiators added to the general gloom before the conference on climate change in Copenhagen. Paul Hohnen and Jeremy Leggett considered any talk about adapting to climate change rather than preventing it, as "an admission of failure." They also argued that a likely last minute "lowest-common-denominator" compromise would only weaken "every nation's commitments to action" and that "the dirty little secret is that not even the most ambitious agreement will meet" the goal of reducing greenhouse gas emissions by 2020 or even 2050. Ironically, they then suggested a "major rethink," the components of which are neither original nor politically feasible.[14]

Yet another recent article by Gideon Rachman in the *Financial Times* of London is even more pessimistic.[15] He contends that the chances of a successful agreement at Copenhagen were "vanishingly small," that the arrival of Obama in the White House "will not be the game-changer that many climate change activists hoped for" (and note the weak bill recently passed by the House of Representatives), and that the proposed deal in which the rich countries "essentially bribe poorer countries to cut emissions and adopt cleaner technologies" and in addition contribute 1% of gross domestice product (GDP) in *additional* foreign aid to help the developing countries participate in the fight against global warming is politically inconceivable. Worse yet, Rachman used information obtained from the climate scientist Oliver Morton to argue that even if a deal had been struck at Copenhagen to cut emissions by 80% by 2050, the promised reductions "seem literally incredible." Finally, while some activists believe a failure at Copenhagen to be catastrophic, "they also know that, even if a deal is reached, it is likely to be feeble and ineffective." Thus the activists are caught on the horns of an agonizing dilemma: if they admit the likelihood of failure, they create or deepen a climate of despair; if they deny failure and press ahead, they will be pursuing an approach that cannot succeed.

To deepen the gloom, at the recent Group of 8 meeting in L'Aquila, Italy (July, 2009) neither China, nor India, nor Brazil indicated any willingness to accept limits on their greenhouse gas emissions. In fact, India indicated that the only limit it would accept is the same per capita amount currently emitted by citizens of the developed world.[16] Whatever the justice of that claim may or may not be, it could be catastrophic for efforts to limit global warming. And when Secretary of State Hillary Clinton visited India shortly thereafter the Indian Environmental Minister seemed almost deliberately insulting in restating India's position aggressively in a joint appearance with Secretary Clinton. Whether this hard line is written in stone or was merely posturing before the meeting in Copenhagen is unclear. What is clear is that flexibility will come at a high price in increased foreign aid and that targets and timetables are only aspirations.

Even if Copenhagen had succeeded in establishing strong targets and timetables for carbon reductions by the major polluters and gradual targets and timetables for the rest of the developing countries, the results might be disappointing. Thus Lomborg notes that even if global emissions were halved by mid-century, it would have a barely measurable effect by the end of the century.[17] While Lomborg may be too pessimistic about the benefits of substantial cuts by 2050, there is bound to be a gap between statements of intentions at this time and performance in the next few decades; it is not without interest that many states have failed to meet the minimal cuts mandated by the Kyoto Protocol and that the growth rate of carbon emissions between 2000 and 2008 was four times greater than the rate from 1990 to 2000.

The Copenhagen Consensus Center, Lomborg's Institute, has commissioned a number of scientific papers that seek to evaluate a number of different forms of climate engineering. One study, for example, has made the case that a relatively small investment in solar radiation management via marine cloud whitening could cancel out the effects of global warming. Consequently, Lomborg and his colleagues argued that the Copenhagen conference was focused on the wrong policies and that geo-engineering would be a far cheaper, far quicker, and far more effective means available to cut emissions. The authors are obviously aware that an extensive research program would be necessary to assess possible side effects and that, even if the tests were successful, they estimated that it could take roughly 5 years before implementation was possible.[18] Still, this and other projects in geo-engineering, if they are feasible and if the side effects seem manageable, *might* provide alternatives to direct carbon reduction (carbon dioxide removal) that are easier to implement, quicker in their effects, and cheaper than other available alternatives.

We should here recall the quip that says "the future is not what it used to be." Given the great uncertainties in our global future, we view any possible geo-engineering options not only as alternatives to our first track or as successors to it if it should prove inadequate. Rather, because of the dangers of failure and policy inertia, we advocate pursuing both tracks *simultaneously* as a form of insurance and as a companion to enhanced research and development of renewables. The premium costs of such insurance will be high, but the costs of doing nothing or being unprepared for some very significant losses could be much greater.

Lastly, some mention is in order of the uncomfortable projection that will face us if nations continue for the next decade or so to resist making changes to greenhouse gas emissions that are now recognized to be essential, and global temperatures continue to rise at an accelerated pace. At some point there is apt to be a strong reaction to their refusal to cooperate and possibly a call for threats of serious economic sanctions, raising the specter of mutually devastating economic nationalism and trade wars. Avoidance of such an awful outcome may yet be one of the strongest arguments for cooperative behavior.

Table 10.1 Summary of priorities: proposed implementation and research over planned time intervals.

Time of Implementation or Research	*Unilateral Goals (Domestic Decisions)*	*Interstate or International Goals (Bilateral or Coalitions)*
First 5 years implementation	Greater use of solar heat Conservation of waste heat Greater appliance efficiency Tighter CAFÉ standards Tax credits for wind and solar use Cap and trade legislation	International agreement on exchange of CCS information EU–US agreement on targets, timetables, monitoring, and enforcement to reduce carbon emissions by 2020
First 5 years research	Evaluate impact of indirect emissions Design smarter electric grid Plan for high speed trains Standardized nuclear plant design Identify deep drilling hazards Find compressed air storage locations Run sequestration trials of CCS	
First decade implementation	Smart electric grid Use centralized nuclear waste disposal Improve mass transit More diesel engine cars Install pumped water storage	US–China agreement on targets, timetables, monitoring, and enforcement to reduce carbon emissions by 2020
First decade research	Develop new high capacity batteries Find best locations for tidal power Study geo-engineering alternatives Extract liquid fuels from algae Improve FT process for liquid fuels Make more efficient PV cells Find wider fuel cell applications	

Continued

Table 10.1 *Continued*

Time of Implementation or Research	*Unilateral Goals (Domestic Decisions)*	*Interstate or International Goals (Bilateral or Coalitions)*
Second decade implementation	Install greater range electric car battery Build new nuclear power plants Install and use more efficient PV cells Extract geothermal energy Build barrages for tidal power Build plant for liquid fuel from algae Operate high-speed intercity train lines Use compressed air energy storage Begin practical carbon capture (CCS)	International agreement on targets, timetables, monitoring, and enforcement to reduce carbon emissions by 2040 International agreement on study of cloud whitening as well as ocean fertilization for plankton growth
Second decade research	Study dye-sensitized solar cells (DSSC) Study ocean changes from CO_2 solution Study reprocessing of nuclear reactor fuel Experiment with geo-engineering via basalt chemical reaction, ocean fertilization for plankton growth, and cloud whitening	
Third decade implementation	Obtain hydrogen from water Geo-engineer climate changes Nuclear fusion feasibility study Nuclear reactors using recycled fuel Run large-scale carbon capture (CCS)	International agreement on fertilizing oceans to grow plankton International agreement on CO_2 disposal into oceans or into basalt deposits
Third decade research	Nuclear fusion research Re-configure cities to shorten commutes Improve water splitting technology Design large fuel cells run on hydrogen	

CCS, carbon capture and sequestration; EU, European Union; FT, Fisher–Tropsch; PV, photovoltaic.

Notes and References

1. Elisabeth Rosenthal, "In Poor Cities Buses May Aid Climate Battle," *New York Times*, July 10, 2009, p. 1.
2. Matthew L. Wald, "What Now for Nuclear Waste?" *Scientific American*, August, 2009, p. 46.
3. Jad Mouawad and Clifford Krauss, "Dark Side of a Natural Gas Boom," *New York Times*, December 8, 2009, p. B1.
4. Hannah Fairfield, "New Scale for Nuclear Power," *New York Times*, December 1, 2009, p. D4.
5. Jad Mouawad and Kate Galbraith, "Study Says Big Impact of the Plug-In Hybrid Will be Decades Away," *New York Times*, December 15, 2009, p. B5.
6. Matthew L. Wald, "Refitted to Bury Emissions, Plant Draws Attention," *New York Times*, September 21, 2009, http://www.nytimes.com/2009/09/22/science/earth/22coal.html.
7. Robert F. Service, "Sunlight in Your Tank," *Science*, Vol. 326, December 11, 2009, p. 1472.
8. Robert F. Service, "New Trick for Splitting Water with Sunlight," *Science*, Vol. 325, September 4, 2009, p. 1200.
9. Jason J. Blackstock and Jane C. S. Long, "The Politics of Geoengineering," *Science*, Vol. 327, January 29, 2010, p. 527.
10. J. J. Blackstock, D. S. Battisti, K. Caldeira, *et al.*, *Climate Engineering Response to Climate Emergencies* (Santa Barbara, CA: Novim, 2009), http://arxiv.org/abs/0907.5140.
11. Gabriele C. Hegerl and Susan Solomon, *Risks of Climate Engineering*, www.scienceexpress.org/6 Aug 2009/page 1/10.1126/science.1178530.
12. Bayden D. Russell and Sean D. Connell, "Honing the Geoengineering Strategy," *Science*, Vol. 327, January 8, 2010, p. 144.
13. Eli Kintisch, "Carbon Sequestration: Should Oceanographers Pump Iron?" *Science*, Vol. 318, November 30, 2007, p. 1368.
14. Paul Hohnen and Jeremy Leggett, "Getting Serious about Climate Change," *International Herald Tribune*, July 13, 2009, p. 8.
15. Gideon Rachman, "Climate Activists are Also in Denial," *Financial Times*, July 28, 2009, p. 9.
16. See "Wanted: Fresh Air," *The Economist*, July 11, 2009, p. 59.
17. See Bjorn Lomborg, *Engineering the Climate: Global Warming's Cheap, Effective Solution*, (Copenhagen: Copenhagen Consensus Center, 2009).
18. J. Eric Bickel and Lee Lane, *An Analysis of Climate Change as a Response to Global Warming*, (Copenhagen: Copenhagen Consensus Center, 2009).

11

Prospects After Copenhagen

For the many scientists and committed environmentalists who believe that global warming is accelerating and that the time left to act effectively to slow or obstruct its progress is rapidly disappearing, the outcome of the Copenhagen conference is an unmitigated disaster. There had been hope on two fronts: first, that the inauguration of the Obama administration would lead to much stronger US leadership and much more willingness to accept firm targets and timetables on emissions, and further that China would demonstrate some leadership on global warming, befitting its new standing in the international system, and agree to meaningful targets and effective monitoring of commitments. In the end each of these major players used the other's failure to act as an excuse for its own inaction, and both hopes were disappointed.

The Challenge of Climate Change: Which Way Now? 1st edition.
By Daniel D. Perlmutter and Robert L. Rothstein.

Did anything of value come from the Copenhagen Accord? There was agreement to hold the temperature rise by the end of the century to less than 2°C and for each country to report plans for voluntary reductions in emissions, but the agreement does not specify amounts and lacks any mode of enforcement. There is no overall goal set for the mid-century. In anticipation of the likely effects of global warming, the developed nations agreed to provide funds for mitigation and adaptation to the extent of $30 billion in the next 3 years, increasing to $100 billion by 2020. There is no provision on who will pay what to whom and when. There was movement on steps to limit deforestation: the developed nations agreed to pay tropical countries to the extent of $3.5 billion to slow forest destruction, but the details of an actual agreement on time, place, and scale of measurements remain to be resolved.

11.1 Costly Failure or Small Success?

The limited results of Copenhagen should not have been entirely unexpected. In the first place, the intensive negotiations that went on for more than a year before the conference had not been able to narrow the policy gaps between different countries and ended in an acknowledgment of failure. There was some earlier movement toward a deal on limiting deforestation, and this did lead to some substantive success at Copenhagen but the US and other key players announced in the months before the conference started that there would be no comprehensive, legally binding agreement and that the most that could be expected was a loose political statement that might someday become binding. This proved to be true.

Furthermore, vesting so many hopes of substantive progress in one huge conference that grouped together 192 countries with vastly different development, environmental and energy contexts, and interpretations of interests, was always excessively optimistic. Whatever such conferences can do in terms of legitimacy and the opportunity to express divergent views, they are the least likely venue to produce what most participants ostensibly wanted: a comprehensive, mutually acceptable, legally binding, verifiable set of firm commitments. Chapters 7 and 8 laid out our doubts about the viability of global conferencing as a decisionmaking forum and Copenhagen has not provided any evidence to suggest that those doubts need to be revised. The organizers for the United Nations (UN) had accredited 45,000 people for a conference hall that held only 15,000, and then spent endless hours arguing about who from the large number of delegations representing the NGO (non-governmental organizations) and civil society networks should or should not be permitted to enter – but a badly organized and badly run conference was only the outward manifestation of much deeper problems.

How can one reach agreement in such circumstances? What has evolved in the UN is a system of intergroup negotiations to avoid the chaos of trying to amalgamate and reach compromises among so many countries. Thus the developing countries meet in the Group of 77 with the goal of devising a coordinated package of proposals, the developed countries have their own group, joined now by most of the states that were once part of the socialist bloc, and China is a "group" of its own. The results are unwieldy packages of demands, some of which are sometimes contradictory and none of which can be traded off without risking group defections. For the developing countries, the easiest, least internally contentious position is simply to demand increased and unconditional transfers of foreign aid, and to threaten to block agreements without such transfers; for the developed countries, the easiest position is to delay, since neither persuasion nor more aggressive postures seemed to produce beneficial outcomes. The group system always seemed necessary in the past to avoid the chaos of amalgamating 192 individual policy preferences, but it has also proved increasingly dysfunctional. Unfortunately, as long as the system of negotiating agreements in grand conferences remained the norm in the UN, the agreements that resulted were bound to be largely rhetorical or symbolic. For more substantive agreements, there was an increasing tendency to rely on smaller groups of more or less "like-minded" states – the Group of 8, the Group of 16, the Group of 20, and so on.

As bad as this situation has been, it got worse at Copenhagen: the groups fell apart, new coalitions could not be constructed, and chaos reigned. The Group of 77 split every which way: China and the other so-called rising powers (India, Brazil, South Africa, sometimes Indonesia) went their own way, the island and low-lying states focused on their own near-term perils, Africa functioned as a loose coalition demanding huge increases in aid, and the remaining Latin American and Asian countries seemed lost in the confusion. Subsequently, there were also disagreements between China, Brazil, and South Africa, which may complicate future negotiations. While Brazil and South Africa supported China during the final negotiations, Brazil subsequently called the agreement "disappointing" and South Africa called the financing commitments "unacceptable."[1]

There was also a sharp split between the United States and the European Union (EU) with the latter much more willing (and able) to offer strong commitments on reducing carbon emissions. In the circumstances, the first 10 or so days of the conference were essentially wasted with the biggest controversy emanating from supposed compromise documents that were apparently deliberately leaked, all of which contained something that one or another group found objectionable.

The absence of leadership, especially by the United States but also by the ostensible UN leadership, was particularly disabling. When President Obama showed up on the last day, apparently hoping to rescue a conference on the verge of disaster, his efforts succeeded in producing only a cosmetic agreement that sought to disguise the fact that none of the conference's original aims had been achieved. The President's description of the Accord as "an unprecedented breakthrough" could only be described as fanciful political hyperbole, or a cynical effort to conceal failure.[2] The situation bordered on the farcical, as the Chinese sent lower ranking officials to two meetings with the President, apparently a deliberate insult, and he had to "invade" a private meeting between the Chinese, the Indians, the Brazilians and perhaps one or two others in order to try and get some agreement that could be passed off as an achievement of sorts.[3] Is it not bizarre that a global conference of nearly 200 countries seeking to achieve agreement on issues of profound importance to every single country should end with a meager 2.5-page document negotiated among five countries with no consultation with the EU, Russia, and the developing countries?

The political agreement that ended the conference was reluctantly, if bitterly, supported by the developing countries and the other "excluded" ones. For the developing countries there was acquiescence only because foreign aid was promised, although the potential gap between what was promised and what would be delivered was worrisome. The deal that included a promise to provide the developing countries $10 billion a year from 2010 to 2012 and $100 billion per annum by 2020 for adaptation, had no details about the sources of this aid or which countries would provide how much of the total. The gap between what the developing countries were demanding and what they were promised was huge, but something was presumably better than nothing. The Chinese refused to accept strict monitoring of their commitment to reduce emissions, which the US had initially insisted was imperative and without which there would be no agreement. Instead, the agreement merely noted that countries would set their own targets and a system for monitoring and reporting progress toward these targets would be established, but without any sanctions for failure or any dispute settlement mechanism. The Chinese also vetoed any effort to set explicit targets and timetables for other countries, perhaps because they feared a precedent that might be set.

The United States was in a weak position to demand more significant concessions on emissions because the President was limited in what he could offer: no more than a 17% cut over 2005 levels by 2020, the minimal amount established by the Waxman–Markey bill passed by the House of Representatives. The developing countries were left to establish and monitor their own emissions, but with some provisions for international help should they choose to ask for it. They were promised an initial $10 billion fund for immediate adaptation, rising to $100 billion by 2020, but without guarantees. As noted earlier, an agreement on deforestation was a very important part of the package (because 18–20% of emissions come from deforestation, mostly in Brazil, Indonesia, and the Philippines) but many of the critical details were left to be negotiated.[4] The commitment to keep the rise in global temperatures below 2°C was a rebuff to the island countries that had lobbied hard for a lower increase of 1.5°C. Stronger commitments for a binding accord and sharp reductions in emissions by 2050 were dropped from the final agreement, largely because of Chinese objections. In any case, nothing in the agreement was legally binding.[5]

As it became increasingly clear both before and during the conference that no comprehensive, binding agreement was possible and that whatever emerged from the last few days of hectic and confused haggling would fall far short of expectations, one question became prominent among many delegates and NGOs: would no deal be better than a weak deal? The apparent assumption of those who preferred no deal was that so resounding a failure, which could not be disguised by vague promises to do something in the future, might energize public support much more effectively than a weak agreement that political leaders could rationalize as a useful first step in an ongoing process. There is probably no clear way to answer this question, but it may in any case have been the wrong question to ask. The more important question is whether the leading nations have learned from the Copenhagen experience that they need to devise a more effective negotiating process in the future, one that inevitably focuses on a smaller group that includes all the major emitters, but with the presumption that they do not use their power and wealth to ignore the needs of the developing countries. We do not yet know the answer to this question, but there are some encouraging signs that other negotiating arenas are already being discussed: the Group of 8 plus, the Group of 16, the Group of 20. Another critical question, which we will seek to address later in this chapter, is whether the domestic and international obstacles to agreement can in any way be diminished? If improvements are not forthcoming, Copenhagen will simply become another in a long string of failures.

Attributing blame for failure or for the very meager results began before the conference even ended. There were, however, several issues mixed together in assessing degrees of responsibility and they probably need to be separated. Specifically, who was most responsible for what went in or was kept out of the final Accord?

Shortly after the conference ended, Ed Miliband, the UK Labour Secretary of State for Energy and Climate Change, specifically accused the Chinese of sabotaging a legally binding agreement to reduce greenhouse gas emissions. A Chinese Foreign Ministry spokeswoman responded sharply that the "British politician" was merely trying to sow discord among the developing countries and create tensions between China and the Group of 77, an attempt "doomed to fail."[6] There is no doubt that as negotiators the Chinese did well at Copenhagen: all their initial positions were intact (including a commitment to cut the overall intensity of carbon emissions in gross domestic product (GDP) by 45%) and they were not obligated to accept any more stringent targets and timetables. But acceptance of such targets and timetables by the developed countries was a central aim of the majority of developing countries because it is at the moment the best means of moderating the rise in temperatures, from which of course the developing countries are likely to be the worst sufferers. Since the Chinese want continued political support from the developing countries and continued access to their natural resources, the charge that they had vetoed provisions in the Accord that would help diminish global warming stung.

Nevertheless, the indictment of China seems fully justified. A much more severe attack on the Chinese position was launched by Mark Lynas, an advisor to one of the delegations present at the final negotiating sessions.[7] He asserts "The truth is this: China wrecked the talks, intentionally humiliated Barack Obama, and insisted on an awful 'deal' so Western leaders would walk away carrying the blame." China insisted, he argued, that all the numbers that mattered (on firm emission cuts by a date certain) be taken out of the deal and replaced by a vague commitment to cut "as soon as possible." Lynas also contends that China got away with this because "it didn't need a deal," the West and the developing countries were desperate for a positive outcome,

President Obama needed to show Congress that he could deliver China, and China had no pressure from below from civil society and a vibrant network of NGOs. The assertion that China did not need a deal is debatable, especially given its current level of environmental deterioration.

In short, the Chinese were the clear "winners" in that they got the deal that they wanted but they were also the clear losers in that Miliband, Lynas and many others have clearly attributed the blame for failure to Chinese obstructionism. The Chinese bargaining position was obviously very strong, since no agreement without strong Chinese assent made much sense, and they were impervious to pressure. For example, threats in the Waxman–Markey environmental legislation to penalize non-compliance with any of its provisions by imposing a carbon border tax on those who failed to commit to carbon reductions did not seem to impress the Chinese (or the Indians), perhaps because such taxes may violate WTO (World Trade Organization) regulations or might unleash a trade war from which all would lose. Still, China's position disappointed more than just the environmental community. There had been hope that with China's rising power and the benefits that it might anticipate from creating a more stable and rule-bound international system would come a rising sense of responsibility to contribute to global order, not merely short-run national interests. To the extent that Copenhagen is a harbinger of China's future behavior, one may wonder whether cooperative outcomes in a variety of areas are now under threat. But before jumping to premature conclusions, it makes sense to try to understand why China chose to act as it did in Copenhagen, a point to which we shall return.

The developing countries were also chided for not bargaining seriously, not including allowance for funds from private investment in their assessments, but simply making a stream of demands for large and unconditional transfers of public resources. The developed countries do not believe that foreign aid expenditures will ever approach the magnitude of developing country demands, while many developing countries resist this argument because they cannot meet the conditions necessary to attract private investment, and they prefer direct government-to-government aid. Then again, they did not have much bargaining leverage, especially when not supported by China and India or when the Group of 77 had trouble reaching a consensus on anything (except more aid).

Matters were not helped by the absurd rhetoric of some Third World delegates. For example, the Sudanese chair of the Group of 77, Lumumba Di-Aping, said the deal was "asking Africa to sign a suicide pact … It's a solution based on values that funneled six million people in Europe into furnaces."[8] His remarks were denounced as "disgusting" and "despicable" by several Western delegates but that they were made at all suggests something about the quality of at least some Third World delegations. In any case, while one must be sympathetic to the plight of countries confronting a potential catastrophe without the material or human resources to cope – and thus in desperate need of external help – absurd rhetoric followed by abject acceptance of the "suicide pact" is not a useful way to seek to influence the negotiations. Ironically, Mr Di-Aping chose not to criticize Sudan's close ally, China, the country most responsible for the failure of the negotiations.

How much real difference will Copenhagen's disappointing outcome make to the effort to develop a coherent, effective, and timely response to climate change? If one is a global warming skeptic or denier, one might applaud the result: no costly and presumably unnecessary policy changes have been imposed on governments. Even some environmentalists, trying to make the most out of a bad situation, have argued that Copenhagen was at least a useful first step in a policy process that will be unending; in fact the process has already resumed in preparation for the next grand conference (in Mexico City in 2010). Of course, if one assumes that the UN is a hopeless venue for serious negotiations, Copenhagen merely reinforces the point.

But even sophisticated analysis that seeks a new approach to the global warming problem after Copenhagen can seemingly lead to a dead end. For example, Bjorn Lomborg has argued that we need a new approach that does not focus on reducing carbon emissions, which he believes has no chance of happening in time to avert disaster, but rather focuses on policies that are technically smarter, more feasible politically, and more efficient economically.[9] Thus we need to increase reliance on renewables by several orders of magnitude because current efforts, even scaled up, would get us less than halfway toward stabilizing carbon emissions by 2050. The only alternative, he believes, is to increase spending on green energy research and development by a factor of fifty.[10] While we are in full agreement about the need to sharply increase research and development spending, the amount that Lomborg suggests is not politically feasible. We must make choices as to which technologies should receive what share of whatever money is available.

We are not necessarily back to ground zero after Copenhagen. Some useful things have been learned about the need to seek a different approach to negotiations, some useful learning curves have been established among at least some developed and developing countries, and there were a few beneficial results at Copenhagen (for example, on deforestation) that might serve as building blocks in the future. How one interprets these points depends on a judgment: how much time do states and the international system have left to react before a rise in temperatures begins to create irreversible patterns of decline and devastation in some parts of our world? Carbon emissions have risen some 20% in the decade since the Kyoto conference, and a number of scientists interviewed by the science editor of the *Financial Times* think that a temperature rise of 3°C is now the best that can be achieved in this century.[11]

Copenhagen has lost us time but not *all* the time we have to act. While global warming is surely occurring, and may be happening at an accelerating pace, there are still too many uncertainties about how rapidly it will occur, how effective responses will be, what costs the responses will entail, and what effects will ensue to simply assume that all is or will soon be lost. Getting an effective policy process established to deal with these issues both domestically and internationally is imperative but frustratingly difficult. Useful ideas about how to overcome the obstacles – political, economic, and psychological – to establishing such a process are difficult to find and then hard to implement. In seeking ways to grapple with these issues, it pays to recall the recent statement from the *The Economist*: "Climate change is the hardest political problem the world has ever had to deal with."[12] The author went on to say that: "It is a prisoner's dilemma, a free-rider problem, and the tragedy of the commons all rolled into one." We find it hard to disagree.

11.2 Reframing the Debate

When an issue or a conflict reaches a stalemate and neither side can impose its will on the other and each can veto whatever the other proposes, it is necessary to seek to reframe the debate. As Lakoff has argued, frames are mental structures that shape the way we see the world: our goals, our plans, our actions, and our evaluations of good or bad outcomes.[13] In effect, frames are about how we conceptualize a situation and, as such, they help us to interpret the world and to organize it coherently. In conflict situations (say, the Middle East or Northern Ireland) or in areas of deep policy disagreement (say, climate change) mutually incompatible frames can become deeply embedded and resistant to change. In such circumstances, progress may require reframing, that is, persuading the other side to alter or adapt its frame so that the mutually incompatible becomes at least partially compatible. Since each frame rationalizes self-interest and seeks to convert the undecided, reframing is never easy. It may require the intervention of a neutral third party, an idea tried many times in a variety of conflicts with an inconsistent pattern of success. The problem may be even more severe domestically in the US, because it is more difficult to imagine who the neutral third party would be and why he or she would be accepted now when levels of party polarization and public opinion divergences seem to have reached near epic proportions. Indeed, as a variety of analysts have noted, change may be so difficult that only a national trauma like 9/11 or a monumental natural disaster like Katrina or the Asian tsunami will generate a sufficient degree of conceptual convergence.

There are a variety of tactics that one might use in an attempt to increase the probability of successful reframing. For example, it makes sense to try to move the debate from the fear of loss (to avoid the "loss aversion" bias that is widely prevalent) to the anticipation of gains. Jonah Lehrer illustrates the point with some familiar examples: twice as many patients opt for surgery when told they have an 80% chance of surviving instead of a 20% chance of dying; and a much higher percentage of people will buy meat when it is labeled 85% lean rather than 15% fat.[14] Thus it might be useful to avoid stressing the fearful consequences of global warming and instead stress the multidimensional benefits to all of responding quickly and effectively. Moreover, there are a number of cognitive biases that one needs to be aware of in seeking to establish a new policy framework. For example, reactive devaluation can generate the belief that a policy is inferior simply because it has been proffered by a hated other (say, a liberal democrat) or the fundamental attribution error can lead to personality-based rather than situation-based explanations of behavior.[15] Conscious awareness of these biases does not, of course, eliminate them but it is a useful first step in containing the potential damage that they can generate.

The initial frames established by the two sides in the global warming debate were quite simple and completely contradictory. The environmental community pointed to the accumulating scientific evidence of global warming, legitimized by increasingly strong warnings from the Intergovernmental Panel on Climate Change (IPCC), and demanded an immediate and strong reaction by the various governments of the world and the international institutional structure. It seemed to be taken on faith that the evidence of warming was strong enough to overwhelm dissent, which was in any case largely dismissed as misguided, ignorant, or the self-interested posturing of the oil and coal industries. Unfortunately, the science of global warming does not suffice by itself because it obviously intersects with too many other issues: economic prosperity, competition for energy supplies, potential water shortages, increasingly arid land, national security, and so on. The much smaller but fervent community of deniers and skeptics either denied that global warming was occurring, or that it was caused by human actions, or that the policies demanded would work or were economically bearable. Both sides soon found that neither frame was adequate to convince enough people of the necessity to do what either wanted done. A frustrating and tenuous stalemate resulted, which did not bother many in the community of deniers but certainly generated some massive anxieties and cries of desperation from many environmentalists. As a result, some on both sides sought to reframe the debate, to create a new narrative that would attract support from the other side.

Some of the efforts to reframe might be described as a tactical retreat or as recognition by political realists of what was needed to get sufficient votes. Thus Senators John Kerry and Barbara Boxer, on the advice of pollsters, refused to use the terms global warming or climate change in their recent bill and instead described it as an effort to put "America back in control of our energy future, reasserting American economic leadership and competitiveness, protecting our families from pollution, and ensuring our national security."[16] The bill, which is not yet passed, also promised "robust border measures" against states that did not commit to cut greenhouse gas emissions. While the bill promised a minimal 20% cut in the level of emissions in 2005, Kerry and Boxer went out of their way to stress that fewer than 2% of American businesses would be affected by these cuts. In short, they reframed the debate in terms of national security, job creation, and protection against presumably unfair competition – and the minimal effects the bill would have on lifestyles or standards of living.

The climate skeptics, deniers, or "realists" (as they now prefer) have also attempted to reframe the debate. They are a very diverse group with very diverse positions, as noted above. One extreme, represented by Professor S. Fred Singer, maintains that the environmentalists "have no evidence. None."[17] However, many in the community of doubters no longer deny the evidence of warming but instead assert that trying to deal with it through sharp reductions in carbon emissions will be too costly and will indeed further impoverish the developing countries (including China and India).[18] This is, of course, a clever reframing tactic because of prevailing economic fears in both the developed and developing countries. It is also true that we will remain dependent on carbon-based fuels for decades to come, that renewables may not receive the massive funding needed to become a viable alternative, and that coal is and is likely to continue to be the cheapest fuel option for China, India, and many other developing countries.

There is no doubt that dealing with climate change will be very expensive, that expenditures on research and development of renewables will have to increase massively and soon, and that adaptation to even existing levels of global warming is beyond the means of developing countries without substantial resource transfers. Because of multiple uncertainties about the future, most estimates of costs are at best rough orders of magnitude.[19] Still, while the numbers appear monumental, they are a small and manageable fraction of global economic output. Moreover, such spending would also have a variety of benefits in terms of jobs, quality of life, and energy security. According to the National Academy of Sciences,[20] the health costs in the US from air pollution alone come to about $120 billion a year. Multiplied by some 200 countries, the costs could be staggering. The poor countries will also obviously benefit if a cap and trade carbon market is established, although private investment will still be necessary.[21] Finally, the economic costs must also be compared to the even greater costs of not spending to avert what could be a linked series of climate catastrophes.[22]

The short-run fear of loss of jobs, income, and homes has dominated the hope of long-term benefits from establishing a new energy economy, and it is difficult in the circumstances to avoid the conclusion that at least in this instance negative reframing has had the upper hand. Perhaps the scientists in the environmental community have been too naïve about how effective scientific evidence will be in a political–ideological debate that is sharply focused on the here and now and with a public as yet more or less indifferent about, or ignorant of, long-term dangers. In 10 or 20 years a new energy economy may provide massive economic benefits and an increase in energy security, but in the next few years the transitional costs could be high, jobs could be lost or relocated, more economic turmoil might be in the offing, and dependence on unreliable and autocratic oil suppliers will still be present. Short-term fear of loss will always outweigh long-run promises of potential benefits, especially if the commitment to intergenerational equity has been diluted by those very fears.

Reframing is again necessary, but one obvious difficulty is that we are in the post-Copenhagen era and there is great uncertainty about the best way to move forward. We are also in the era of linked, multidimensional crises that can feed on each other and use up resources that are scarce. Thus, to have a chance at success any new frame of reference must emphasize this very complex set of circumstances: whatever can and cannot be done internationally and whatever other countries do or do not do domestically, it is imperative for every country, but especially the major polluters now and in the next few decades, to establish a policy process that starts producing effective policies as rapidly as possible, policies that can be the basis for international cooperation should that become possible and that provide some degree of protection if cooperation continues to flounder. National action alone cannot deal with global warming, but it is a crucial first step.

Suggestions about how to deal with global warming post-Copenhagen would be much easier if we could discern a set of trends or developments pointing in a clear direction. We want in the next section to discuss some of the major roadblocks to effective policymaking on global warming: Chinese international policies, public opinion and other political constraints in the US, and the attitudes and policies adopted by most of the developing countries. It is not possible or sensible, however, to discuss these issues in isolation from other trends and developments that are occurring, or may be perceived as occurring, in the international arena. Simply put, neither China, nor the US, nor the developing countries should make policy in a vacuum. What follows thus provides a brief list of favorable and unfavorable trends and developments post-Copenhagen, with the strong qualification that some of these trends or developments might be quickly reversed, if, for example, the world economy moves out of recession. The list is meant primarily to serve as a framework or background for the political discussion that follows.

11.3 The Good News and the Bad News

First, the bad news. The failure of Copenhagen is obviously the most salient negative development, although its effects may have been diluted by awareness well before the conference that a binding agreement on targets and timetables was not going to be possible. The very muted political and public response to the failure may, however, be even more ominous: without pressure from below the likelihood of strong responses to global warming is very low. In any case, the limited public support for energy and environmental policies is so crucial that we shall return to it again.

China was the short-run winner at Copenhagen but its long-term strategy is obscure since it also suffers severely from environmental depredations but does not seem willing to take strong action to avert even worse outcomes or to cooperate to reach binding international agreement. It was a disappointment that China does not seem to have adjusted its policies and has not taken a leadership role as it has rapidly risen in the hierarchy of states. Its policies are essentially rather like a revival of mercantilism: the pursuit of narrow national interests, the capture of foreign natural resources, the effort to win export markets at all costs, and the refusal to accept binding commitments that limit national actions. The stability of the international system could be undermined if China's policies force others to follow suit, thus generating a dangerous cycle of conflict. International systems function badly if leaders are not willing to sacrifice some short-run gains in order to spread security and prosperity to the weaker states. In effect, states should and will pursue national interests but should also avoid policies that threaten international stability and prosperity. A well-known example of such behavior is that of the US after World War II, when the financial and trading systems of both defeated Japan and Germany were supported so that they could more quickly become stronger allies against the Soviets. Unfortunately, as yet, China does not see its responsibilities in this sense; it seems to act solely for the benefit of its short-term interests.[23]

In the absence of a binding agreement at Copenhagen to begin cutting emissions rapidly, a recent report by the International Energy Agency (IEA) in Paris[24] has predicted soaring energy consumption in the next few decades, resulting in a catastrophic rise in temperatures of up to 3°C (5.4°F). While the current recession has facilitated a 3% drop in emissions this year, the IEA predicts a 40% rise by 2030 (half from China alone, much of the rest from other developing countries) and a 76% rise in electricity demand in the same period, much of it coming from burning coal. The report concluded that each year of delay in reaching a binding agreement will eventually require an additional $500 billion per annum to cut emissions. Further in the wrong direction, there has also been a sharp drop in investments in renewable energy by the large oil companies and others.[25] Seeing this, countries trying to assess the medium-term availability of oil and gas supplies and the prospects for replacement by renewables will be more hesitant to change, thereby fostering competitive efforts to control supplies (as in China) and a reduced willingness to cooperate or to take long-run shared interests into account.

There is yet another potentially negative factor. Keeping the rise in temperatures below 2°C (3.6°F) has always seemed a daunting goal because the requisite reductions in per capita emissions are so demanding. It would require limiting overall per capita emissions to about 2 tons, but current US levels are about 20–24 tons per capita, Europe about 10 tons, and China about 5 tons and rising rapidly. Getting these numbers down in any relatively short-term period – say the next 10–20 years – is very demanding, not to say impossible if China and the other developing countries continue to refuse to accept binding cuts. Moreover, the means of achieving these cuts – government regulations, carbon pricing, and various subsidies – are as yet uncertain in their effects and the anticipated costs of acting to stay under a 2°C rise would require something of the order of a trillion dollars per year for an extended period, of which about half would be spent in developing countries.[26]

Whatever the accuracy of these estimates of cost or of the effects of a temperature rise above 2°C, the problem is that setting such demanding goals in the short run with such severe costs may generate fatalism, or indifference, or disbelief, not to mention fears about negative economic consequences. Nicholas Stern has estimated the costs of reaching such goals as about 1% of GDP per annum but, given the inefficiencies of the policy process, this may be an underestimate.[27] The amounts are not inconceivable in either economic or technical terms but they are probably politically infeasible in the current political and economic environment in the US and many other countries, at least until public opinion has shifted and until congressional sentiments on climate change have altered.

The failure to persuade the developing countries that their negotiating position is dysfunctional is another negative factor. These countries have focused on the arguments that they have not created the problem of global warming, that they are most vulnerable to its effects, and that they are owed substantial compensation for past sins of the developed countries and to avert catastrophic damage if temperatures continue to rise. They also insist that per capita emissions, which are obviously very low in poor countries, are the proper standard, not the absolute amounts, which are beginning to grow in many countries. Some of these arguments are valid and there are in any case both moral and practical reasons to provide the developing countries with as much help as possible regardless of how guilt is apportioned.

The arguments that the developing countries have made and are continuing to make are emotionally powerful and reflect deeply felt grievances. But they are almost beside the point. Whoever is at fault for the severe problems that global warming has created and is creating, the dilemma that we now face is that reparations for past sins, justified or not, are a diversion from what needs to be done: every state needs to act, regardless of who got us into this situation or is likely to keep us there. The old cliché appears to be true: if we do not hang together, we shall all hang separately. Global warming ignores borders and emissions from developing countries are rising sharply and will constitute the vast majority of future emissions. Proclaiming that development needs must and will take precedence over systemic needs may be politically, economically, and emotionally evocative arguments, but they are also dangerous arguments. Pleas for massive resource transfers, though they may be mutually beneficial, will fail unless the developing countries are also willing to take serious, monitored actions to rein in their own emissions. All are too important a part of the problem to not be part of the solution: free riding will no longer suffice.

Given the failures of so many aid projects in the past, substantial resource transfers will become much more likely only if the developing countries show that they are serious about implementing policies to reduce emissions and not if they refuse to act until they receive reparations. In addition, their demand that they receive aid without conditions of transparency and accountability fortifies the suspicion that much of the aid will be lost to corruption or diverted to other needs that have little to do with curbing global warming. Unfortunately there was almost no sign at Copenhagen that the developing countries were willing or able to rethink positions that were deeply embedded, emotionally satisfying, and undemanding in terms of policy changes. The profound vulnerabilities they confront should focus attention on immediate practicalities, but continued dominance in so many cases by weak and incompetent governments tends to keep the focus on staying in power, not investing heavily in projects to adapt to current and emergent threats.

We are all aware that domestic politics may be the most difficult obstacle to overcome if an effective policy process on global warming is to be established. President Obama's performance at Copenhagen may have been disappointing to many environmental activists, but it was vastly superior to the position of the Bush administration. It appears that he went as far as he could go without starting a domestic struggle with Congress. Recent political events suggest that President Obama may be even more sharply constrained after the 2010 mid-term elections. The decision by several prominent Democrats to retire is symptomatic of deeper potential problems. As Zeleny and

Nagourney argue, "a conservative push against the president's ambitious agenda, a sluggish recovery from the deep recession and an outbreak of angry populism" may combine to produce large gains for the Republicans in this year's congressional races.[28] Moreover, the sharp and nasty polarization that has come to characterize political debate in the US implies that policies will be discussed and supported or rejected not on their merits but simply on the basis of who is supporting them or identified with them. This obviously bodes ill for the future of environmental legislation and for the prospects for success at the Mexico City conference next year. But the mood of the electorate is fickle, the elections are still months away, and the anticipation of negative outcomes may be premature; all is not yet lost.

To put the picture in balance, there is some good news too on the technology and business fronts: some things have happened that might make the prospects for an effective and efficient policy response more likely, if they come to fruition before the negative trends and developments undermine any attempts to cooperate. For one, the new technique to tap previously inaccessible supplies of natural gas from shale has the potential to sharply increase global reserves of the cleanest fossil fuel. Estimates of future production are enormous, stretching from a conservative increase of 20% in the world's known reserves to 160%, justifying Daniel Yergin's statement that this new method of gas extraction "is the biggest energy innovation of the decade."[29] There are two crucial points here: as more of this gas is produced and replaces coal, global warming could be sharply reduced; and it could also reduce dependence on unreliable oil and gas producers – Russian, Venezuela – a great political and economic gain for Europe and other importers. It should be emphasized, however, that massive up-front investment will be required, potentially crowding out other investment needs, and that it will probably take more than a decade before these new supplies are widely available.

Another promising development is the emergence of competition among energy producers, and there are newly divergent interests among natural gas producers, oil producers, and the coal industry. Electrical utilities are finding disagreements among themselves over the use of coal or wind power, and of course the renewable power industry is seeking an edge against all the other energy sources.[30] In addition, it is of some importance that in the US a number of powerful companies, such as General Electric and Pacific Gas, have withdrawn from the Chamber of Commerce in protest at the latter's opposition to strong legislation on global warming. Further developments in this direction are uncertain, strongly depending on the future price of oil and, in turn, its effect on investments in renewables, relations among oil producers, and economic growth rates.

On the eve of the Copenhagen conference, the US Environmental Protection Agency announced that greenhouse gases posed a danger to human health and the environment, thus opening the possibility of controlling emissions by governmental regulatory action, thereby by-passing the need for direct congressional action.[31] The preferred route to cut emissions is through the legislative process, however, since legislative action is inherently more powerful and more difficult to overturn. Moreover, depending on the size and composition of the vote for such action, it is likely to seem more legitimate and less partisan. It should be added here that a new dimension has been introduced into the governmental role in the climate change debate. In recent actions, two federal appeals courts in the US ruled that climate change lawsuits could go forward.[32] In Alaska, Connecticut, and Mississippi property owners are bringing suit against energy producers, claiming that they were injured by global warming and seeking damages. It is not clear how these cases will be handled, but they are being compared by some to the early suits against tobacco and asbestos companies. If in fact these suits move through the courts as did the earlier tobacco and asbestos cases, the US judiciary may turn out to be more effective in creating change than either the executive or the legislature.

We have already noted a few other positive developments such as the (tentative) deal on deforestation at Copenhagen and the appearance of a new administration in the US that does not simply deny global warming, ignore the need for cooperation, or believe it can go it alone. But even the good will of the Obama administration has not sufficed because the administration is constrained by what it can get through Congress in a very fraught period.

One must admit in assessing this list of potentially good or bad trends and developments that at the moment the negative strongly outweigh the positive and can effectively block a timely response to global warming. The positive trends and developments are weaker and more tentative, and they may not matter much if China remains resistant to legally binding action and the US is limited by Congress and the lack of public support.

The stakes of the debate on climate change are obviously enormous. If the worst forecasts come to pass, the international system could become dangerously chaotic and unstable, but we believe we yet have 10 or 20 years to avert the worst. The lesser changes may still create massive problems of adaptation and adjustment, but they are potentially manageable. Everything is in play:

the fate of the planet, of nations, of citizens rich and poor. The supreme issue is responsiveness, that is, how well and how quickly states and international institutions respond to challenges that are complex, multidimensional, and unending. However, states and international institutions have not responded effectively because the threats from global warming do not seem as immediate, as clear, or as compelling as other costly threats and because the deniers and skeptics have framed the debate to rationalize inaction. There seem to be three matters that will determine whether we succeed or fail in our policy responses: (i) the policies of China, (ii) the stance of the Third World, and (iii) the inability to get public and congressional support for strong policies in the US. These are the subjects of the next sections.

11.4 The China Problem

China began to invest heavily in new energy technology early in this new century and then redoubled its efforts in 2006, with strong emphasis on jumping ahead in solar and wind technology. Reflecting President Hu Jintao's instruction that China "must seize preemptive opportunities in the new round of the global energy revolution,"[33] the nation began spending as much on green technology as on its military budget, about $100 billion per annum. Its stimulus package targeted 38% on green technology, and it aims to increase solar and wind generation massively by 2020. Given China's protection of new industries, its supply of cheap labor, the massive number of engineering graduates pouring out of its universities, and its closed domestic market, the move to invest heavily in renewable energy technologies shows promise to dominate export markets and perhaps earn massive and growing returns.[34]

Evidently the Chinese leadership, which includes many individuals with technical training, takes global warming seriously, and yet their strategy at Copenhagen was to refuse to accept legally binding targets and timetables for the reduction of emissions. Indeed they even blocked approval for such targets and timetables for other states. It appears that China's domestic strategy, emphasizing rapid growth with strong mercantilist inclinations, overshadows any leanings toward international cooperation. It is fitting then to ask what – if anything – can be done to alter Chinese behavior so that it puts greater weight on cooperative, mutually beneficial outcomes, not mere competitive advantage.[35]

An entry to this subject is first to ask what China's vulnerabilities are if they continue on their present course. The most important of these may be China's own internal environmental threats which will be exacerbated when climate change follows increased burning of fossil fuels. China is undergoing a sharp movement of population from the countryside to urban areas, a change that will increase energy demand, dependence on foreign oil supplies, and concomitant problems of energy security. They may be expected to see air pollution, land degradation, and a combination of flooding in some areas and drought in others as patterns of rainfall begin to alter. Such happenings would at the very least demand significant increases in health and recovery expenditures, and might eventually create rising levels of domestic dissent and instability.[36]

Moreover, as an emerging power in the world, China needs outside markets and resources. They may find both more difficult to obtain if they are seen to always act unilaterally without concern for world opinion. China benefits hugely from the opportunities available in an open international trading system, but the system can start to unravel if its trading partners do not also benefit. One can already see signs of resistance: legislative actions in both the US and France to impose tariffs on exports from countries that do not accept binding commitments to cut emissions, and populist demands to take even more severe actions against China for "stealing" jobs unfairly. Furthermore, there are growing complaints in some developing countries that China is taking their resources too cheaply and bringing in Chinese staff rather than relying on native workers, warning signs that too narrow a focus on self-interest may become self-defeating.

The Chinese leadership needs to be convinced that it is in its own interests to rethink its positions. China no longer has the luxury of making its policies in isolation from the rest of the world. Precisely because it is so powerful, its actions redound throughout the international economic and political systems. When it refused at Copenhagen to accept restraints on its behavior, the refusal inflicted severe damage on its ostensible allies in the developing world.[37] Further controversy has arisen from China's behavior in maintaining an artificially cheap currency and amassing a huge stash of over $2 trillion of currency reserves, while signing many contracts to acquire natural resources from developing countries. In doing as it has to sustain very high growth rates, China is meeting the desires of its people for higher standards of living, but at the cost of inviting rising domestic dissent from a citizenry becoming better educated and more demanding politically. In short, China's leadership must be persuaded that its policies need to be altered in order to avoid confronting an increasingly hostile international community and an increasingly disaffected domestic citizenry.

The hope is that China may in the near future become open to a "grand bargain," and some suggested terms of a possible deal have been floated.[38] For a deal to be made China would have to agree to accept legally binding targets and timetables to reduce emissions of greenhouse gases (GHGs) and to have such steps monitored. China's commitments would need be implemented over a relatively short period of time, say 5 years, before more severe reductions in emissions were mandated, and sanctions for non-compliance would become punitive only over a 10–20-year period. In exchange for accepting such terms, China could be guaranteed that no restrictions would be arbitrarily imposed on its exports, that research and development of new energy technologies would be shared with the costs distributed fairly, that special provisions would be established for equitable treatment of other developing countries, and that no country that developed a technology first would seek to earn monopoly rents from that competitive advantage. Money matters could be left for the World Bank to control, or, following Gallagher's idea, a Carbon Mitigation Fund could be created to finance the incremental costs of adopting low carbon technologies in China and developing countries.

Why would the Chinese leadership accept such a deal? Whether they are based on anticipated rising domestic dissent arising from increasingly unequal income distribution or from unfavorable external developments such as a deeper recession, trade wars, or nationalist reactions in the Third World, predictions that China's "bubble" is about to burst are at least premature. They may, however, become plausible in the next 5–10 years. Thomas Friedman has written about "a political class focused on addressing its real problems."[39] A leadership that successfully devised its own domestic "grand bargain" and that navigated the transition from communism to a hybrid system of tight political control and a carefully controlled opening to the world economy should be far-sighted enough to see the need to begin adjusting its policies to take account of its leadership responsibilities and to avert populist reactions that might threaten the viability of that domestic bargain.

China is also concerned with being treated with respect and having its new status recognized and institutionalized. Continuing to defy a growing international consensus on global warming could begin to threaten or undermine this status. Lastly, it should be added that there are also some areas of potentially overlapping mutual interests that could make a deal more attractive.

One such area would involve the US focusing on research and financing, and the Chinese focusing on their comparative advantage in mass production. How to share benefits of any arrangement along these lines and how to prevent the Chinese from selfishly appropriating the new technology would have to be carefully negotiated.

Would such negotiations with China bear fruit? A provisional answer is to be found in Chou en Lai's famous response to a question asking him to assess the French Revolution: "It's too soon to tell."

11.5 Third World Dilemmas

In spite of their differences in levels of development, resource endowments, and political and ideological orientations, the poor and weak developing countries have since the middle years of the twentieth century attempted to maintain unity in their positions vis-à-vis the First (developed) World. They have insisted that the problems of the Third World are the legacy of colonialism exacerbated by unfair terms of trade for exports of natural resources, and unfair access to the markets of the developed countries; they have demanded increasing amounts of foreign aid, preferably without conditions such as transparency and accountability. Further they have argued that because they are not responsible for the current levels of GHG emissions they should not accept any limitations on their efforts to grow rapidly via the same carbon-intensive model that worked for the developed countries.

On a superficial level nothing much seemed to have changed at Copenhagen. Demands for foreign aid were vastly increased, blame for existing levels of GHG emissions was attributed to the developed countries, and compensation was asked for. But, on another level, some important things had changed that challenged Third World unity. While rhetorical unity was intact, substantive unity was not. As noted earlier the island states, the African states, the Latin Americas, and the "emerging powers" of India, Brazil, Indonesia, and South Africa were in pursuit of their own immediate interests, not all of which were necessarily compatible. There was an air of near desperation implicit in some developing country demands for massive, immediate, and unconditional resource transfers. In the past aid was helpful, but except for a favored few, was not given in sufficient amounts to make much difference, particularly when an important part of the aid was wasted on corruption, incompetence, and excessive military expenditures.[40] Now, however, as some poor countries begin to confront the magnitude of the challenges they already face from environmental deterioration, development failures, deficient or non-existent infrastructure, and insufficient technical expertise, a sense of how bad conditions might soon get seems to be growing.[41] Perhaps this implies that even some of the ruling elites may finally be recognizing that they too will be heavily implicated if adaptations to global warming are inadequate. If so, they may move more seriously to plan for a future circumstance when they will accept aid attached to strong performance criteria. Finally, whatever the reasons were that justified foreign aid in the past, the developed countries now need the developing countries to make a significant contribution to lowering future emissions of GHGs. We are, in effect, linked together in a classic global commons problem that may mean sharp increases in foreign aid will be both necessary and possible.

Do these changes suggest that a deal with the Third World is now more likely than it has been? Can we influence the developing countries to abandon their obsession with ancient resentments, and to refocus on what they need to do to protect themselves against increases in global temperatures? The divisions within the Third World that surfaced at Copenhagen do suggest that some developing countries feel the need to move beyond standard Group of 77 posturing. They need help quickly and they may be willing to accept some conditions on whatever aid is on offer to start getting effective policies established. These countries should be offered both technical assistance for policy planning and conditional aid to begin implementing such plans. It needs to be made clear by the US, the EU, and the various aid organizations that similar offers are on the table for any country seeking help, and that the aim is not to establish a closed club of favored recipients but rather to acknowledge and allow for the differences that exist within the Third World bloc. It makes sense to start cooperating immediately with those who are aware of the need to do so. If this option were on the table and taken seriously by many of the poorer countries, China and the other emerging powers might also be induced to become more cooperative.

In addition it should be recognized that many of the smaller developing countries are fearful of moving without the support of China, India, Brazil, Indonesia, and South Africa, but while there are some overlaps in their goals these larger powers are primarily focused on achieving their national interests, not Group of 77 interests. The policy positions taken at Copenhagen by China, India, and Brazil are illustrative: they made private deals with the US and the EU without much consultation with the Group of 77; nor did they pay particular attention to the Group of 77's special concerns. The aim should be to make clear that it is sensible for the smaller and poorer countries to seek new options that protect or enhance their interests, rather than allow China to pursue its own interests behind an ostensible commitment to the interests of the developing countries.

It is understood that any offer to the smaller and poorer developing countries will require a multistage approach, because most of these countries are in no condition to make strong or legally binding commitments at this time. Timing is not crucial, however, since these nations are not *yet* major polluters, and a decade will probably pass before they become important contributors to global warming. What is needed at the moment is technical assistance and enough foreign aid or private investment to begin implementing adaptation policies. This fits the standard UN formulation of "common but differentiated responsibilities" for the developing countries, and as Jing Cao has advocated, this means that they will take on additional responsibilities for reducing emissions but only gradually.[42] Cao also suggested a formula to assess individual and state responsibility for GHG emissions, which would facilitate agreement among donors about appropriate shares of the financing needed to help these countries.

Some developing countries will probably get substantial help in the next 3–5 years, not as much as they need and certainly not what they think they deserve, but enough to begin protecting themselves against the worst and nearest environmental threats. Others will get very little and may even be added to the list of failed states. To expect more in this political and economic environment is neither sensible nor prudent in view of the many other legitimate uses for these resources and the strong political resistance to demands for massive aid increases. The hope is that increasing numbers of developing countries will read the signs, heed the warnings, and either develop alternative growth strategies or become pragmatic meliorists until a time when more aid becomes possible.

11.6 Polarized Politics

President Obama was severely limited in Copenhagen by the existing terms of the Waxman–Markey House bill and the Kerry–Boxer Senate bill, which had not yet been acted on. Both bills were weaker in their GHG restrictions than the several European plans already in effect. Furthermore, no political leader in a democracy can get too far in front of his constituents, and recent electoral outcomes in Virginia, Massachusetts, and New Jersey are clear warnings that limit the Obama administration's space for maneuver on environmental issues. In response, Senators Graham and Kerry "turned their sights to a more modest package of climate and energy measures," without any effective cap and trade scheme to put a price on carbon emissions.[43]

A Pew Research Center poll which appeared 2 months before the Copenhagen conference also provided evidence of the limited support in the US for strong government policies on global warming.[44] There has been an across the board decline in the belief that there is solid evidence of global warming: from 75% in April 2008 to 53% in October, with an even sharper decline among Republicans from 62% in 2007, to 49% in 2008, to 35% in 2009. In addition, only 36% of the respondents attributed global warming to human activity. The decline has been especially severe among Republicans but Democratic support also declined, if less significantly. Only 14% of the Republicans polled thought global warming was a serious problem in contrast to 49% of Democrats. In January 2009 global warming was the lowest-rated priority for both Republicans and Independents and ranked sixteenth among twenty issues for Democrats. This was, of course, at a time of great economic turbulence as well as there being serious concerns about Iraq, Afghanistan, and health care reform, but it does indicate how difficult it is to generate strong support for serious responses to the threat of global warming or to create any kind of bottom-up pressure on political leaders to act.

A subsequent report from the Pew Research Center[45] sought to shed some light on why public belief in the evidence of global warming had declined so sharply. The state of the economy was primary, but the Pew Report also suggested several other factors: an unusually cool summer, fear that the effort to limit emissions would increase prices and lead to more job losses, and the influence of the most-watched television news and most-listened to talk radio. A Gallup environmental survey in March 2009, cited in the Pew Report, reached essentially similar conclusions. In short, although opinions were quite fluid, many people were aware of global warming and many thought it was or would become a serious problem, but it was not a problem of sufficient salience to override all other issues: the worst effects of global warming did not seem likely to occur for several decades and were most likely to have immediately severe consequences for "others" – that is, the poor and distant.[46] People with a limited interest in a problem that still seems essentially long term are not likely to be roused to action or to think about altering comfortable and conventional views by more scientific studies or warnings of looming disaster.

Moreover, because the outlook of administrations and Congress can be altered by every election, stakeholders fear quite sensibly that the possibilities of non-compliance will increase and that demands for renegotiation of terms will emerge. Such fears can arise especially when trust in government and elected representatives is very low and when the parties and the electorate are deeply polarized. To assess the extent of the current polarization in Congress, the political scientist Barbara Sinclair has documented the rise of threatened or actual filibusters since the 1960s: such actions affected 8% of major legislation in the 1960s, 27% in the 1980s, and 70% after Democrats took control of Congress in 2006.[47] Cass Sunstein, a professor at Harvard Law School, has provided another illustration of how selective polarization has occurred. In analyzing a large number of internet sites he found that each side in a debate tends to look largely at other sites with a similar point of view. Thus proponents of one view become more extreme after communicating only or primarily with proponents of the same view: group polarization is the primary result and mutual compromise becomes harder and harder to achieve.[48] Widespread internet use, therefore, seems to confirm biases: users seek out those with the same view, which serves to solidify that view, but also makes the position more extreme and more firmly held.

There is abundant evidence of this pattern of reinforcement in the global warming debate. For example, Senator Inhofe from Oklahoma, who described global warming as "the greatest hoax ever perpetrated on the American people," cited only the small group of scientists or advocates who support his views and totally ignored the large body of scientific evidence that refutes them.[49] Reasoned debate on the merits is the first casualty of this kind of selective perception and deep polarization: each side ignores or derides the other, attributes opposing positions to greed or ignorance, and sees no need to challenge its own certainties. Political constraints on effective policymaking are not limited to the US. A well-known political columnist in England recently attacked the unwillingness of the leadership to take political risks to advance the green agenda, and criticized some parts of the UK press for denying the existence of global warming. She also noted that polling data on Tory candidates in the next general election indicated that they wanted less action on energy and the environment and a cutback on foreign aid.[50]

In these difficult circumstances strong political leadership on energy and the environment will become possible only if public opinion begins to shift and to demand effective policies from Congress and the President in the US and the relevant authorities in other countries. The missing link in persuading our political leaders to act has been the absence of effective bottom-up pressure or awareness of the possibility that their hold on power might be threatened by an aggrieved public who have become aware of the dangers of doing nothing.[51] Is it possible to generate such pressure, even without a disaster that is clearly linked to global warming? A positive answer will be possible only if the debate on global warming can be reframed more effectively than simply linking it to jobs and national security. The latter are surely crucial matters, but we need to search elsewhere to understand why support for strong action against global warming is weak and why strong contrary evidence can be so easily ignored.

The issues need to be reframed to impact decisionmaking and voter behavior by affecting both emotions and reasoning. Thomas Friedman has taken a step in this direction by his efforts to alert the public to the dangers of global warming and continued dependence on petro-autocrats. He has asked what the deniers and skeptics must believe to justify their position. He answers that someone holding such a view must also believe that the population will not increase by 2.5 billion people by 2050, that it is good for the US to remain dependent on potentially hostile or unstable oil exporters, and that the poor people of the world will not seek to enjoy the lifestyles we in the developed world have enjoyed for many decades.[52] This is a useful way of illuminating some of the weaknesses in the analyses of the community of deniers and skeptics, but it does not tell us enough about why they hold such views and why they are so resistant to altering them.

If we are to answer these questions we need to focus on the impact of both reason and emotion on decisionmaking and political behavior. The debate between environmentalists and doubters has deteriorated sharply: too often neither listens to the other nor seeks to understand the other's position. How can one break down this wall and begin to discover enough shared values and interests to permit some mutually acceptable and effective compromises to emerge? Part of the answer can be derived from the field of political psychology, which goes some way toward explaining how partisans of any view can rationalize a refusal to consider divergent views or to reconsider their own. Examining their findings may suggest ways to reframe the debate on global warming more effectively.

Drew Westen[53] has opined that "the political brain is an emotional brain," that faced with threatening or discordant information, partisans of any political persuasion are thus likely to "reason" to emotionally biased conclusions. Lehrer and Westen both maintain that in fact we are all rationalizers.[54] This point is crucial in the present context, because it helps us to understand why more information does not necessarily diminish bias; "voters tend to assimilate only those facts that confirm what they already believe."[55] Cognitive dissonance should arise if new information contradicts firmly held beliefs, but not if the new information is derided or ignored.

George Marcus and his colleagues have argued that emotions not only affect what we are feeling "but also how and about what we think, and what we do."[56] Emotions can lead people away from habitual responses and toward new policies or new ways of thinking. These emotions are most effectively stimulated by high levels of anxiety if they are accompanied by the feeling that improvement or a way out of a dilemma is possible; if anxiety is low, debate and discussion do not change minds or elicit new patterns of behavior.[57] This is again crucial in our context because the narrative espoused by deniers and skeptics is designed to lower anxiety and to suggest that large and costly changes in behavior are unnecessary.

Given the abundant evidence that people are past masters at rationalizing their beliefs and that they resist change even in the face of contradictory evidence, what hope is there of reframing the debate on global warming in a fashion that is persuasive to majorities on both sides? The narrative of the skeptics and deniers has argued that there is time to adapt and adjust gradually to whatever changes do occur and that the anxieties of the environmental community are misplaced or self-interested special pleading. They have also stressed repeatedly that a quick switch to a new energy economy will raise costs substantially. All of this generates an inappropriate perception of certainty and supports the tendency to overvalue immediate gains and to undervalue the future: immediate losses loom larger than potential gains.[58] There is no easy formula to overcome or diminish the strength of this narrative; it is too comfortable a fit and too easy a regimen for "true believers."

Philip Tetlock[59] has studied the fallibility of expert political judgments and provides a powerful warning about the dangers of what he calls the "sin of certainty." Premature certainty and selective simplification are the factors that he finds to be mainly responsible for the misjudgment of the experts. Tetlock is not questioning scientific judgments based on evidence, but many of the opinions in the debate about global warming are not scientific, and Tetlock's warnings seem to be both timely and relevant. He notes also that one of the main characteristics of the more accurate forecasters was their willingness to consider deviant views and to question their own beliefs and assumptions.

In spite of much inflammatory rhetoric, there are a few issues in the global warming debate that offer some possibilities for uncontroversial cooperation. We need to use them to our advantage. In this spirit we propose two approaches that might change the way in which the debate is carried on and begin to generate some useful patterns of communication. The first thrust would focus on improving local capacities for adaptation, an area of concern that could form the basis for immediate efforts at cooperation. The second emphasis, directly reflecting our earlier discussion of the need to appeal to *both* emotions and reason, attempts to reframe the debate in terms of intergenerational equity. In seeking to explain why a present generation is willing to spend on investments that will primarily benefit later generations, it may be observed that there is a kind of tacit bargain between generations: we invest to help future generations because prior generations have done the same for us. When this bargain breaks down and short-run selfishness becomes the rule, age-related resentments arise and social stability may be threatened.

Even though it will probably be the earliest hit by crises generated by global warming and the most severely affected, the local level has not received enough attention. The floundering local responses to the Asian tsunami, to hurricane Katrina, and to the events of 9/11 illustrate that this neglect is a mistake. There is something of a pattern in these cases: the national leadership tends to be weak and unaware of what is going on locally, local staff are badly trained and forced to improvise a response without proper equipment or supplies, communications tend to break down, and the responsible agencies are largely concerned with deflecting blame. Adequate advance planning should remedy or at least diminish some of these deficiencies.

Since the developing countries lack the financial resources or the technical manpower to handle large-scale disasters, the initial use of the aid funds promised at Copenhagen should be to mandate the development of national plans to respond quickly to disasters. Technical help in devising these plans, training for local staff, and the provision of adequate equipment for communications, emergency medical services, and evacuation plans should be included. For the developed countries, which do have the necessary resources and skills, the national authorities should demand that local governments prepare beforehand and be authorized to act quickly without waiting for national directions. Whereas national authorities are most important in the pre-crisis and post-crisis stages, it is the local authorities who will have to cope in the crisis phase. Because there is general agreement on the benefits of a more effective first response, and because the costs of preparation are relatively modest, improving the local capacity to respond ought to be relatively uncontroversial. The vast range of potential catastrophes implies that institutions at all levels must be flexible, adapted to the task at hand, and even partially redundant. They must be prepared to deal with low probability events having great consequences.

Earlier attempts to reframe the debate failed because the frame articulated by skeptics and deniers responded effectively to the current concerns and needs of the public, and because the proponents of a strong policy response to global warming, fearful of political defeat, have felt the need to respond in the language of their opponents. The new formulation must be cognitively persuasive but, as our previous discussion indicated, it must also be emotionally evocative: reason alone can only provide the means but not the ends we want to pursue.[60] Appealing to both emotions and reason is not a new idea, and there are many illustrations of the political power of emotional appeals.[61] Although the point has been lost in the noise of our present polarized, polemical, and populist political arena, the issues of global warming and energy security ought not to be inherently partisan. What shared values could overcome or diminish these divisions?

Our sense of commitment to future generations and our sense of empathy for those most in need have been impaired by our current circumstances, but they have not been completely destroyed as evidenced by the outpouring of aid for the victims of the Asian tsunami and the earthquake in Haiti. Such feelings need to be resurrected and strengthened along two avenues. First, both sides should be asked, but especially the skeptics and deniers, to imagine that they might be wrong. Suppose that only some part of the worst forecasts about climate change prove to be true: the consequences could still be disastrous for some areas of the world and those in the richer world will not be able to isolate themselves, except at great moral and practical cost, from some of the worst effects. And even if the forecasts of temperature increases were to be proved wrong, there would still be enormous benefits from investments in renewable energy and a reduction in energy dependence on the petro-dictators. In short, even if you are a doubter, it makes eminent sense to take out some insurance against the possibility that your doubts are misplaced. Reducing worry and anxiety is an important part of the reason for paying the costs of insurance. It can also be a crucial element in efforts to change minds and policies. Arguing, as some doubters have, that we need not be anxious is to put an extraordinarily large bet on the likelihood that the majority of the world's scientists do not know what they are talking about.

The other component of this approach is also in the form of an appeal to the community of disbelievers. The appeal rests on the judgment that despite many differences there are shared values in our commitment to future generations and empathy for those less able to respond to disasters. Do we not owe to future generations a good faith effort to leave them a planet that is livable on? No matter what one's political beliefs are or what estimates of short-run electoral gains may exist, does it not make more sense to fight the partisan battles on other grounds and to make a shared commitment now to our children and grandchildren? Political power should be used for more lofty goals than mere partisan advantage. Are you comfortable living with the notion that your grandchildren may remember you for failing to provide them with a chance to live lives as full and prosperous as ours, or failing to take out insurance against the possibility that you might be wrong? Are you absolutely certain that all the warnings of looming dangers by so many scientists from so many scientific disciplines are wrong? Such certainty in the face of so much contrary evidence is frightening and may be judged as irresponsible by our descendents.

Notes and References

1. Fiona Harvey, "Climate Change Alliance Crumbling," *Financial Times*, December 22, 2009 (on-line edition).
2. John M. Broder, "5 Nations Forge Pact on Climate; Goals Go Unmet," *New York Times*, December 19, 2009, p. 1.
3. On the apparent snub to the President and his entrance into a meeting to which he had not been invited, see ibid, p. A10.
4. On the difficulties of negotiating the details, see "Touch Wood," *The Economist*, December 19, 2009, p. 112.
5. John M. Broder, "5 Nations Forge Pact on Climate; Goals Go Unmet," op. cit., has a very useful summary of what is in and what is not in the final agreement. Also useful is Andrew C. Revkin and John M. Broder, "Grudging Accord on Climate, Along with Plenty of Discord," *New York Times*, December 20, 2009, p. 1 and 4. The full text can be found in "Copenhagen Accord Final Text," *The Guardian*, December 21, 2009 (on-line edition).
6. See Ed Miliband, "The Road from Copenhagen," *The Guardian*, December 20, 2009 (on-line edition). For the Chinese response, see Andrew Jacobs, "Chinese

and British Officials Tangle in Testy Exchange over Climate Change," *New York Times*, December 23, 2009, p. 14.

7. Mark Lynas, "How Do I Know China Wrecked the Copenhagen Deal? I was in the Room," *The Guardian*, December 23, 2009, p. 10.
8. Quoted in John Vidal and Jonathan Watts, "Copenhagen Closes in Weak Deal that Poor Threaten to Reject," *The Guardian*, December 19, 2009 (on-line edition).
9. See Bjorn Lomborg, "We Should Change Tack on Climate after Copenhagen," *Financial Times*, December 22, 2009 (on-line edition); also see Peter Viebahn, Manfred Fishedick, and Daniel Vallentin, "Carbon Capture and Storage," pp. 99–102 in *State of the World 2009: Into a Warming World* (New York: W. W. Norton and Company, 2009).
10. Bjorn Lomborg, "We Should Change Tack on Climate after Copenhagen," ibid.
11. Clive Cookson, "Hopes for New Order from Climate Chaos," *Financial Times*, December 17, 2009 (on-line edition).
12. "Getting Warmer," *The Economist*, December 5, 2009, p. 4.
13. George Lakoff, *Don't Think of an Elephant! Know Your Values and Frame the Debate* (White River Junction, VT: Chelsea Green Publishing Company, 2004), p. x and 4ff.
14. Jonah Lehrer, *How We Decide* (New York: Houghton Mifflin Harcourt, 2009), pp. 106–7.
15. George Lakoff, *The Political Mind* (New York: Viking, 2008), pp. 224–9.
16. Quoted in Suzanne Goldenberg, "Democrats' Bill Pushes Senate to Act over Climate Change," *The Guardian*, October 1, 2009, p. 23.
17. See Tom Zeller, Jr., "And in this Corner, Climate Doubters," *New York Times*, December 10, 2009, p. A6.
18. See, for example, the view of Lord Monckton, ibid.
19. Krugman cites the results of a study by the Congressional Budget Office that indicates restricting emissions would have a modest effect on projected average annual growth rates of GDP both domestically and internationally (roughly between 0.03% and 0.09% per annum). See Paul Krugman, "Green Economics: How We can Afford to Tackle Climate Change," *New York Times Magazine*, April 11, 2010, p. 39.
20. Matthew L. Wald, "Fossil Fuels' Hidden Costs is in Billions, Study Says," *New York Times*, November 11, 2009, p. A16.
21. A reasoned critique of conservative opposition to either cap and trade or a carbon tax can be found in Robert H. Frank, "Of Individual Liberty and Cap and Trade," *New York Times*, January 10, 2010, p. B7.
22. John M. Broder, "Climate Deal Likely to Bear Big Price Tag," *New York Times*, December 9, 2009, p. 1 and A10.
23. And China shows few signs of altering its behavior. Its demand for power from oil and gas "has led to the largest six-month increase in tonnage of human

generated gases ever by a single country." This is in the immediate post-Copenhagen period. See Keith Bradsher, "In China, Soaring Energy Appetite Threatens Emission Goals," *New York Times*, May 7, 2010, p. B1.

24. See Jad Mouawad, "Gloomy Energy Report Sets the Stage for Climate Negotiations," *New York Times*, November 11, 2009, p. B4.
25. See "Getting Warmer," *The Economist*, December 5, 2009, p. 8. It should be noted that China is an exception to the cutback in spending on renewables.
26. For the numbers, citing an International Energy Agency report, see "Getting Warmer," ibid., p. 20.
27. Ibid., p. 8.
28. Jeff Zeleny and Adam Nagourney, "Party is Shaken as 2 Democrats Choose to Quit," *New York Times*, December 7, 2010, p. 1 and A20.
29. Clifford Krauss, "New Way to Tap Gas may Expand Global Supplies," *New York Times*, October 10, 2009, p. 1 and 3.
30. John Broder and Jad Mouawad, "Energy Firms Find No Unity on Climate Bill," *New York Times*, October 19, 2009, p. 1 and A18.
31. John M. Broder, "Greenhouse Gases Imperil Health, E.P.A.," *New York Times*, December 7, 2009, p. 1.
32. John Schwartz, "Courts Emerging as Battlefield for Fights Over Climate Change," *New York Times*, January 27, 2010, p. 1.
33. Quoted in Evan Osnos, "Green Giant," *The New Yorker*, December 21, 2009, p. 54.
34. Robert Kennedy, Jr., "The New (Green) Arms Race," *Outreach* (Copenhagen: Stakeholders Forum), December 7, 2009; also see Martin Jacques, *When China Rules the World: the End of the Western World and the Birth of a New Global Order* (New York: Penguin Press, 2009).
35. For an excellent overview of the issues raised by China's rise, see C. Fred Bergsten, Charles Freeman, Nicholas R. Lardy, and Derek J. Mitchell, *China's Rise: Challenges and Opportunities* (Washington, DC: Peterson Institute for International Economics and the Center for Strategic and International Studies, 2009).
36. Yingling Liu, "A Chinese Perspective on Climate and Energy," pp. 84–7, and Jennifer Wallace, "The Security Dimensions of Climate Change," pp. 63–6, in *State of the World 2009: Into a Warming World* (New York: W. W. Norton and Company for the Worldwatch Institute, 2009).
37. On China's currency policies, see Paul Krugman, "Chinese New Year," *New York Times*, January 1, 2010, p. A25.
38. Kelly Sims Gallagher, *Breaking the Climate Impasse with China: a Global Solution*, Discussion Paper No. 09-32 (Cambridge, MA: Harvard Project on International Climate Agreements, Belfer Center for Science and International Affairs, Harvard Kennedy School, November 2009).

39. Thomas L. Friedman, "Is China the Next Enron?" *New York Times*, January 1, 2010, p. A27.
40. William Easterly, *The White Man's Burden: Why the West's Efforts to Aid the Rest have done So Much Ill and So Little Good* (New York: Penguin Press, 2006).
41. Personal communication from a delegate to the Copenhagen conference, who talked to many Third World delegates and commentators.
42. Jing Cao, *Reconciling Human Development and Climate Protection: Perspectives from Developing Countries on Post-2012 International Climate Change Policy*, Discussion Paper No. 08-25 (Cambridge, MA: Harvard Project on International Climate Agreements, Belfer Center for Science and International Affairs, Harvard Kennedy School, December 2008).
43. See John M. Broder and Clifford Krauss, "Advocates of Climate Bill Scale Down their Goals," *New York Times*, January 27, 2010, p. A4.
44. *Fewer Americans See Solid Evidence of Global Warming* (Pew Research Center Publications, October 22, 2009). All the polling data in the paragraph comes from this publication's on-line edition.
45. *Searching for Clues in the Global Warming Puzzle* (Pew Research Center for the People and the Press, October 27, 2009) (on-line edition).
46. Marten Scheffer, Francis Westley, and William Brock, "Slow Response of Societies to New Problems: Causes and Costs," *Ecosystems*, Vol. 6, No. 5, August, 2003, p. 495.
47. Cited in Paul Krugman, "A Dangerous Dysfunction," *New York Times*, December 21, 2009, p. A29.
48. Cass Sunstein, *Going to Extremes: How Like Minds Unite and Divide* (New York: Oxford University Press, 2009).
49. The quote and other materials are from a speech on the Senate floor on January 4, 2005, reproduced on the Senator's internet site.
50. Polly Toynbee, "Gutless, Yes. But the Planet's Future is No Priority of Ours," *The Guardian*, December 18, 2009 (on-line edition).
51. Since the young have been more concerned with environmental issues than the rest of the population, there have been some tentative suggestions that the Republicans might alter their hostility toward environmental policies in order to broaden their electoral base. Unfortunately, there is as yet no sign of this. It is possible, however, that the ongoing oil spill disaster in the Gulf of Mexico *might* finally begin to break the wall of resistance to strong environmental policies. Much will depend on how long the crisis continues and bad its near-term effects are.
52. Thomas L. Friedman, "What They Really Believe," *New York Times*, November 18, 2009, p. A31; also Thomas L. Friedman, *Hot, Flat, and Crowded* (New York: Farrar, Straus and Giroux, 2008).

53. Drew Westen, *The Political Brain – the Role of Emotion in Deciding the Fate of the Nation* (New York: Public Affairs, 2007), p. xv (italics in original).
54. Jonah Lehrer, *How We Decide* (New York: Houghton Mifflin Harcourt, 2009), p. 205; Drew Westen, ibid., p. xi.
55. Johan Lehrer, *How We Decide*, ibid, p. 206.
56. George E. Marcus, W. Russell Neuman, and Michael MacKuen, *Affective Intelligence and Political Judgment* (Chicago: University of Chicago Press, 2000), p. 38.
57. Richard Nadeau, Richard Niemi, and Timothy Amato, "Emotions, Issue Importance, and Political Learning," *American Journal of Political Science*, Vol. 39, No. 3 (August 1995), pp. 558–74.
58. See Jonah Lehrer, *How We Decide*, op. cit., pp. 77–81, and George Lakoff, *The Political Mind*, op. cit., p. 228.
59. Philip Tetlock, *Expert Political Judgment: How Good Is It? How Can We Know?* (Princeton, NJ: Princeton University Press, 2005).
60. See Drew Westen, *The Political Brain*, op. cit., p. 133ff.
61. The previously cited works by D. Westem, G. Lakoff, J. Lehrer, and G. E. Marcus are illustrative.

Index

Note: page numbers in *italics* refer to figures; page numbers in **bold** refer to tables

The Challenge of Climate Change: Which Way Now? 1st edition.
By Daniel D. Perlmutter and Robert L. Rothstein.